COMPUTER SCIENCE, TECHNOLOGY AND APPLICATIONS

PARALLEL COMPUTING USING REVERSIBLE QUANTUM SYSTOLIC NETWORKS AND THEIR SUPER-FAST ARRAY ENTANGLEMENT

COMPUTER SCIENCE, TECHNOLOGY AND APPLICATIONS

Additional books in this series can be found on Nova's website under the Series tab.

COMPUTER SCIENCE, TECHNOLOGY AND APPLICATIONS

PARALLEL COMPUTING USING REVERSIBLE QUANTUM SYSTOLIC NETWORKS AND THEIR SUPER-FAST ARRAY ENTANGLEMENT

ANAS N. AL-RABADI

Nova Science Publishers, Inc.
New York

Copyright ©2011 by Nova Science Publishers, Inc.

All rights reserved. No part of this book may be reproduced, stored in a retrieval system or transmitted in any form or by any means: electronic, electrostatic, magnetic, tape, mechanical photocopying, recording or otherwise without the written permission of the Publisher.

For permission to use material from this book please contact us:
Telephone 631-231-7269; Fax 631-231-8175
Web Site: http://www.novapublishers.com

NOTICE TO THE READER

The Publisher has taken reasonable care in the preparation of this book, but makes no expressed or implied warranty of any kind and assumes no responsibility for any errors or omissions. No liability is assumed for incidental or consequential damages in connection with or arising out of information contained in this book. The Publisher shall not be liable for any special, consequential, or exemplary damages resulting, in whole or in part, from the readers' use of, or reliance upon, this material. Any parts of this book based on government reports are so indicated and copyright is claimed for those parts to the extent applicable to compilations of such works.

Independent verification should be sought for any data, advice or recommendations contained in this book. In addition, no responsibility is assumed by the publisher for any injury and/or damage to persons or property arising from any methods, products, instructions, ideas or otherwise contained in this publication.

This publication is designed to provide accurate and authoritative information with regard to the subject matter covered herein. It is sold with the clear understanding that the Publisher is not engaged in rendering legal or any other professional services. If legal or any other expert assistance is required, the services of a competent person should be sought. FROM A DECLARATION OF PARTICIPANTS JOINTLY ADOPTED BY A COMMITTEE OF THE AMERICAN BAR ASSOCIATION AND A COMMITTEE OF PUBLISHERS.

Additional color graphics may be available in the e-book version of this book.

LIBRARY OF CONGRESS CATALOGING-IN-PUBLICATION DATA

Al-Rabadi, Anas N., 1973-
  Parallel computing using reversible quantum systolic networks and their super-fast array entanglement / author, Anas N. Al-Rabadi.
      p. cm.
  Includes bibliographical references and index.
  ISBN 978-1-61122-741-3 (softcover)
 1. Parallel processing (Electronic computers) 2. Systolic array circuits.
 3. Quantum electronics. 4. Quantum computers. I. Title.
  QA76.58.A528 2010
  004'.35--dc22
                         2010041592

Published by Nova Science Publishers, Inc. ✝ New York

Contents

Preface		vii
About the Author		ix
Chapter 1	Introduction	1
Chapter 2	Fundamentals of Systolic Arrays	5
Chapter 3	Reversible Circuits and Systems	15
Chapter 4	Reversible Systolic Networks	25
Chapter 5	Quantum Systolic Networks	35
Chapter 6	Conclusions and Future Work	47
References		51
Index		59

PREFACE

In quantum computing, and because all of the states of the quantum system can exist simultaneously, all of the paths of the quantum computations tree from the root to the leaves occur in parallel and only after measurement a single path will be observed as the whole system's composite state will collapse into that single path (i.e., state). From computation perspective, each path in the tree of quantum computing is a single processing, and thus a massive computational parallelism exists with massive number of calculations performed simultaneously and the path superposition will collapse after measurement into a single path, where the path with the highest probability has the highest probability to be measured. New type of m-ary systolic arrays called reversible systolic arrays is introduced in this research. The m-ary quantum systolic networks are also introduced. A systolic array is an example of a single-instruction multiple-data machine in which each processing element performs a single simple operation. Systolic devices provide inexpensive but massive calculation power, and are cost-effective, high-performance, and special-purpose systems that have wide range of implementations such as in solving several regular and compute-bound problems containing repetitive multiple operations on large arrays of data. Similar to the classical case, information in a reversible and quantum systolic circuit flows between cells in a pipelined fashion, and communication with the outside world occurs only at the boundary cells. Since basic PEs that are used in the construction of arithmetic systolic networks are the add-multiply cells, the results that are introduced in this work are general and apply to a wide spectrum of add-multiply-based systolic networks. Since the reduction of power consumption is a main specification for the circuit design in future technologies, such as in quantum computing, the main features of several future technologies will include reversibility. Consequently, the new systolic networks will play an important task in the design of future circuits and systems that consume minimal power. It is also shown that the new systolic arrays maintain the high level of regularity while exhibiting the new fundamental reversibility and quantum superposition properties. These new properties will be essential in performing super-fast arithmetic-intensive calculations that are basic in several future implementations such as in multi-dimensional quantum signal processing.

ABOUT THE AUTHOR

Anas N. Al-Rabadi is currently an Associate Professor in the Computer Engineering Department at The University of Jordan. Dr. Al-Rabadi received his Ph.D. in Advanced Logic Synthesis and Computer Design from the Electrical and Computer Engineering Department at Portland State University in 2002, received his M.Sc. in Power Electronics Systems Design and Feedback Control Systems Design from the Electrical and Computer Engineering Department at Portland State University in 1998, and was a Research Faculty at the Office of Graduate Studies and Research (OGSR) at Portland State University. He is the author of the first comprehensive graduate-level book and the first published title on Reversible Logic Synthesis, *Reversible Logic Synthesis*: *From Fundamentals to Quantum*

Computing, (Springer-Verlag, 2004). Currently, Dr. Al-Rabadi has published more than 100 scholarly articles in international journals and conferences, in addition to a nanotechnology patent registered in the USPTO and several published Book chapters, where his current research includes computer organization and architecture, parallel and distributed computing, systolic architectures, regular circuits and systems, reversible logic, quantum computing, many-valued logic, soft computing, optical computing, reconstructability analysis, signal and image processing, design for testability, nanotechnology, robotics, optimal and robust control, fractals, contact of rough surfaces, error-control coding, computer networks, embedded systems, and residue arithmetic. He is currently an editor of several international journals, and a member of IEEE, ACM, Sigma Xi, Tau Beta Pi, Eta Kappa Nu, OSA, SIAM, SPIE, APS, INNS, IIIS, and ASEE.

Chapter 1

INTRODUCTION

In quantum computing, and because all systems' states can occur at the same time, all of the paths of the tree in quantum computing occur in parallel, and only after measurement, a single path will be observed as the whole system's superimposed state will collapse into that single state, thus, from computational point of view, each path in the tree of quantum computation is a single computation, and therefore a massive computational parallelism occurs with large number of calculations performed simultaneously where the path with the highest probability has the highest probability to be measured [1,66].

Around the year 2020, the anticipated breakdown of the empirical observation of Moore's law, stating that the number of transistors that can be placed inexpensively on an integrated circuit doubles approximately every two years, will occur. Therefore, quantum computing will hopefully play an increasingly important role in building more compact and less power consuming computers [1,7,23,30,65,66,82]. Due to this fact, and because all quantum computer gates (i.e., fundamental building blocks) should be reversible [1,2,3,7,9,18,23,30,57,65,66,72,82], reversible computing will have an increasingly more existence in the future design of regular, compact, and universal circuits. (k, k) reversible circuits are circuits that have the same number of inputs k and outputs k and are one-to-one mappings between vectors of inputs and outputs, thus the vector of input states can be always uniquely reconstructed from the vector of output states [1,9,18,23,57,66]. As was proven [57], it is a necessary, but not sufficient, condition for not dissipating power in any physical circuit that all system circuits must be built using fully reversible logical components. For this reason, different technologies have been studied to implement reversible logic in hardware like adiabatic CMOS [78], optical [72], and quantum [1,66].

Fully reversible digital systems will greatly reduce the power consumption (theoretically eliminate) through three conditions: (1) logical reversibility: the vector of input states can always be uniquely reconstructed from the vector of output states, (2) physical reversibility: the physical switch operates bidirectionally backwards as well as forwards, and (3) using ideal-like switches that have no parasitic resistances.

Simple and regular interconnections lead to cheap implementations and high densities, and higher density implies both higher performance and lower overhead for support components [1,13,64,71,87], where it has been shown that multi-dimensional pipelining plus multi-processing at each stage of a pipeline can lead to the best-possible computing performance [42,51-53,81,86]. Systolic systems provide inexpensive and massive parallel computation power, and provide a model of computation which captures the concepts of pipelining, parallelism, and interconnection structures which has been implemented in wide range of applications [11,14-16,20-22,24-29,31-39,42,43,48,52-56,58-63,69,70,73,74,83,84,86,88-100], and provide a model of computation for studying parallel algorithms for VLSI that takes into account issues of I/O, control, and inter-processor communication. In a systolic system, multi-dimensional pipelining can overlap I/O with computation to ensure high throughput with no extra control logic is required [52-54]. In systolic circuits, communication paths inherently require more space and energy than processing elements (PEs) do, and communication among processors is performed through fixed data paths, where these paths have simple and regular geometries. It has been shown that data flow patterns in systolic systems are fundamental in matrix computations [42,52-54,74]: e.g., the two-way flow on the linearly connected network is common to both matrix-vector multiplication and the solution of triangular linear systems, and the three-way flow on the hexagonally mesh-connected network is common to both matrix multiplication and the LU-decomposition.

The main contribution of this work is the new synthesis method that implements m-ary functions reversibly and through the realization of m-ary reversible functional expansions using parallel-based reversible systolic networks and the corresponding parallel-based entangled quantum systolic networks as illustrated in Figure 1. The new reversible logic synthesis method possesses high-level of regularity in addition to the preservation of the reversibility property. In addition to the reversibility property, the extension of the new reversible systolic arrays to the quantum domain will result in the emergence of the quantum superposition property which is responsible of the exponential speedups of the computational processes in the quantum domain [1,66]. One of the advantages of the use of new families of quantum systolic networks is their potential utilization in low-power circuit designs for digital signal processing applications in analogy to the role of classical systolic networks in non-adiabatic VLSI circuits [41,53], where also the use of quantum circuit technology speeds up the computational process due to the existence of the quantum computational parallelism.

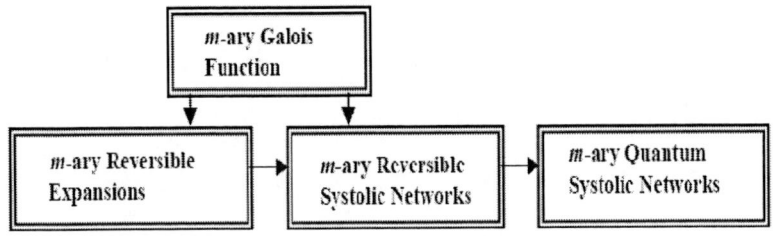

Figure 1. The implementation of m-ary (i.e., many-valued) functions using reversible functional expansions and reversible systolic networks and their corresponding quantum systolic networks.

Basic background on systolic arrays is given in Chapter 2. Fundamentals of reversible logic are presented in Chapter 3. New architectures of reversible systolic arrays are introduced in Chapter 4. Quantum realizations and computations of the new m-ary reversible systolic architectures are introduced in Chapter 5. Conclusions and future work are presented in Chapter 6.

Chapter 2

FUNDAMENTALS OF SYSTOLIC ARRAYS

Systolic arrays are examples of parallel-based single-instruction multiple-data (SIMD) machines in which each processing element (PE) is only capable of performing a single simple operation [6,11,14-16,20-22,24-29,31-39,42,43,45,47-49,52-56,58-63,69,70,73,74,80,83,84,86,88-100]. In a systolic computing system, data flows from the host through the array in a rhythmic fashion and computations are synchronized by a global clock signal where data items are pumped out from a memory [42,52-54]. In systolic arrays, the function of the processor is analogous to that of the heart; every processor (i.e., processing element) regularly pumps data in and out, each time performing a short computation, so that regular data flow is kept up in the network [42,52-54].

The power behind the systolic arrays comes from the way in which the data flows between the processing elements. Typically a systolic array is capable of performing a single operation such as matrix multiplication or inversion. They are thus special purpose machines used mainly in dedicated equipment and not in general purpose computers. Several problems can be solved using systolic arrays [42,52-54]. These include convolution, finite impulse response (FIR) filtering, infinite impulse response (IIR) filtering, discrete Fourier transform (DFT), interpolation, matrix-vector and matrix-matrix multiplication, and the solution of linear systems [79]. Systolic arrays may be used to solve many regular problems containing repetitive operations on large arrays of data, but still not suited to general purpose computers, although with the development of more general programmable PEs, programmable systolic arrays may be possible [20,24,25,29].

A systolic system consists of a set of interconnected cells, each capable of performing some simple operation. Properly designed parallel structures that need to communicate only with their nearest neighbors will gain the most from very-large-scale-integration (VLSI) technologies, because valuable time is usually lost when modules that are far apart must communicate. Systolic arrays are cost-effective, high-performance, special-purpose systems that have wide range of applications. Because simple, regular communication and control structures have substantial advantages over complicated ones in design and implementation, cells in a systolic system are typically interconnected to form a systolic array or a systolic tree. Information in a systolic system flows between cells in a pipelined fashion, and communication with the outside world occurs only at the "boundary" cells; only those cells on the array boundaries may be I/O ports for the system. The basic function of a systolic array is achieved by replacing a single PE with an array of PEs, and a higher computation throughput can be achieved without increasing the memory bandwidth [42,52-54]. As mentioned previously, the function of the memory is analogous to that of the heart; it pulses data from memory through the array of PEs. The central point of this approach is to ensure that once a data item is brought out from the memory it can be used effectively at each cell it passes, and this is possible in a wide class of compute-bound computations where multiple operations are performed on each data item in a repetitive manner. The gain in processing speed in millions of operations per second (MOPS) can be justified with the fact that the number of pipeline stages has been increased n-times (equals n-PEs). Other advantages of systolic arrays include modular expansionability, simple and regular data and control flows, use of simple and uniform cells, elimination of global broadcasting, limited fan-in, and fast response time.

Linearly connected, orthogonally connected, and hexagonally connected PEs are examples of mesh-connected systolic arrays [42,52,54]. Various systolic configurations have been shown along with their potential usage in performing computations [42,52,54]: 1D linear arrays are suitable for convolution, FIR filter and discrete Fourier transform (DFT); 2D square arrays are suitable for dynamic programming and graph algorithms; 2D hexagonal arrays are suitable for matrix arithmetic and DFT; Trees are suitable for searching algorithms; and triangular arrays are suitable for inversion of triangular matrix and formal language recognition. The I/O is the bottleneck in systolic systems where the major problem with a systolic array is still in its I/O barrier, and the globally structured systolic array can speed-up the computations only if the I/O bandwidth is high [42,52-54].

2.1. KUNG HEXAGONAL SYSTOLIC ARRAY

Band matrices are important since several scientific and engineering computations involve such matrices, and since a dense matrix can be viewed as a band matrix having the maximum-possible bandwidth [1,10,40,41,44,46,66,67,68,75,85]. Each pulsation of a systolic array consists of the following operations [42,52-54]: (1) shift and (2) multiply and add. Basic processing cells that are used in the construction of systolic arithmetic arrays are the add-multiply cells. This kind of cells has the three inputs $\{a, b, c\}$ and the three outputs are $\{a = a, b = b, d = c + a * b\}$. One can assume six interface registers are attached at the I/O ports of a processing cell. All registers are clocked for synchronous transfer of data among adjacent cells. The add-multiply operation is needed in performing the inner product of two vectors, matrix-matrix multiplication, matrix inversion, and LU decomposition of a dense matrix. Hexagonally connected processors (i.e., processing elements) can optimally perform matrix multiplication [42,52-54]. In a hexagonal systolic array, three data streams flow through the array in a pipelined fashion. One can follow the operation of the 2D hexagonal systolic array by studying the data flow by moving transparencies of the band matrices over the network (cf. Figure 2). Multiplication of band matrices [A] and [B], [A] · [B] = [C], and the associated definition of bandwidth is shown as follows [42,52-54]:

$$\begin{bmatrix} a_{11} & a_{12} & 0 & 0 & 0 \\ a_{21} & a_{22} & a_{23} & 0 & 0 \\ a_{31} & a_{32} & a_{33} & a_{34} & 0 \\ 0 & a_{42} & a_{43} & a_{44} & a_{45} \\ 0 & 0 & a_{53} & a_{54} & a_{55} \end{bmatrix} \cdot \begin{bmatrix} b_{11} & b_{12} & b_{13} & 0 & 0 \\ b_{21} & b_{22} & b_{23} & b_{24} & 0 \\ 0 & b_{32} & b_{33} & b_{34} & b_{35} \\ 0 & 0 & b_{43} & b_{44} & b_{45} \\ 0 & 0 & 0 & b_{54} & b_{55} \end{bmatrix} = \begin{bmatrix} c_{11} & c_{12} & c_{13} & c_{14} & 0 \\ c_{21} & c_{22} & c_{23} & c_{24} & c_{25} \\ c_{31} & c_{32} & c_{33} & c_{34} & c_{35} \\ c_{41} & c_{42} & c_{43} & c_{44} & c_{45} \\ 0 & c_{52} & c_{53} & c_{54} & c_{55} \end{bmatrix}$$

Band_1 Band_2 Band_3

Where: bandwidth$_1$: $w_1 = 3 + 2 - 1 = 4$, bandwidth$_2$: $w_2 = 2 + 3 - 1 = 4$, and bandwidth$_3$: $w_3 = w_1 + w_2 - 1 = 4 + 4 - 1 = 7$. Figure 2 shows the two-dimensional hexagonal systolic array [42,52-54] that implements the operation of multiplying two band matrices [A] and [B].

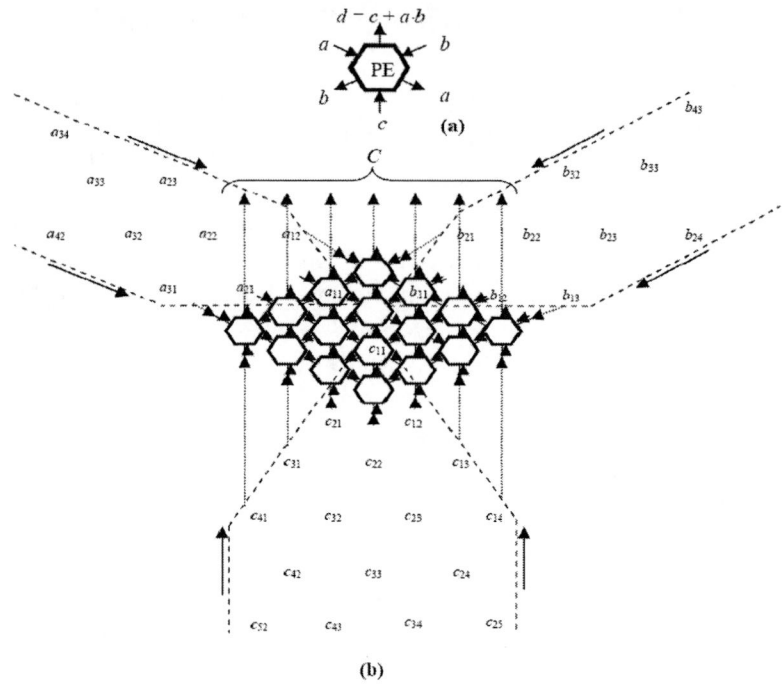

Figure 2. Hexagonal two-dimensional systolic array: (a) Kung cell and (b) Kung systolic array.

Each front data level in the three data flow streams in Figure 2 is called a wave front, the initial values of the input [C] array elements (from the lower side in Figure 2) are all zeros, and the final (resulting) values of the output [C] array elements (from the upper side in Figure 2) are obtained as follows:

$c_{11} = a_{11}b_{11} + a_{12}b_{21}$, $c_{21} = a_{21}b_{11} + a_{22}b_{21}$, $c_{31} = a_{31}b_{11} + a_{32}b_{21}$, $c_{41} = a_{42}b_{21}$, $c_{51} = 0$, $c_{12} = a_{11}b_{12} + a_{12}b_{22}$,
$c_{22} = a_{21}b_{12} + a_{22}b_{22} + a_{23}b_{32}$, $c_{32} = a_{31}b_{12} + a_{32}b_{22} + a_{33}b_{32}$, $c_{42} = a_{42}b_{22} + a_{43}b_{32}$, $c_{52} = a_{53}b_{32}$,
$c_{13} = a_{11}b_{13} + a_{12}b_{23}$, $c_{23} = a_{21}b_{13} + a_{22}b_{23} + a_{23}b_{33}$, $c_{33} = a_{31}b_{13} + a_{32}b_{23} + a_{33}b_{33} + a_{34}b_{43}$,
$c_{43} = a_{42}b_{23} + a_{43}b_{33} + a_{44}b_{43}$, $c_{53} = a_{53}b_{33} + a_{54}b_{43}$, $c_{14} = a_{12}b_{24}$, $c_{24} = a_{22}b_{24} + a_{23}b_{34}$,
$c_{34} = a_{32}b_{24} + a_{33}b_{34} + a_{34}b_{44}$, $c_{44} = a_{42}b_{24} + a_{43}b_{34} + a_{44}b_{44} + a_{45}b_{54}$, $c_{54} = a_{53}b_{34} + a_{54}b_{44} + a_{55}b_{54}$,
$c_{15} = 0$, $c_{25} = a_{23}b_{35}$, $c_{35} = a_{33}b_{35} + a_{34}b_{45}$, $c_{45} = a_{43}b_{35} + a_{44}b_{45} + a_{45}b_{55}$, $c_{55} = a_{53}b_{35} + a_{54}b_{45} + a_{55}b_{55}$.

The topological distribution of the processing elements in the systolic structure, shown in Figure 2, is obtained as follows: # PEs in top-left = w_1 = 4, # PEs in top-right = w_2 = 4, # PEs in bottom = w_3 = 7, and the total # PEs = 4 · 4 = 16.

2.2. LINEAR CONVOLUTION SYSTOLIC ARRAY

The convolution problem can be viewed as matrix-vector multiplication where the matrix is a triangular Toeplitz matrix [44,76]. Another example of the application of Toeplitz-type matrices is the p-tap finite impulse response (FIR) filter which is a matrix-vector multiplication where the matrix is a band upper triangular Toeplitz matrix with bandwidth $w = p$ [44,67,68,75]. Using linearly connected systolic array of size n, both the convolution of two n-vectors and the n-point DFT can be computed in $O(n)$ units of time rather than $O(n \log n)$ as required by the sequential fast Fourier transform (FFT) algorithm [42,44,52,67,68,75].

As an example of the linear convolution, let $u(n)$ and $w(n)$ be causal sequences and each is of finite length N, then the linear convolution of $u(n)$ and $w(n)$ is a causal sequence computed as [44,67,68,75]:

$$y(n) = u(n) * w(n) = \sum_{k=0}^{N-1} u(k)w(n-k), \quad n = 0,1,2,...,(2N-2).$$

The linear convolution of two vectors:

$$\{\vec{a} = \{a_0, a_1, a_2, a_3, a_4\}, \vec{x} = \{x_0, x_1, x_2, x_3, x_4\}\},$$

can be represented in a matrix form called Toeplitz matrix as:

$$\begin{bmatrix} a_0 & 0 & 0 & 0 & 0 \\ a_1 & a_0 & 0 & 0 & 0 \\ a_2 & a_1 & a_0 & 0 & 0 \\ a_3 & a_2 & a_1 & a_0 & 0 \\ a_4 & a_3 & a_2 & a_1 & a_0 \end{bmatrix} \begin{bmatrix} x_0 \\ x_1 \\ x_2 \\ x_3 \\ x_4 \end{bmatrix} = \begin{bmatrix} b_0 \\ b_1 \\ b_2 \\ b_3 \\ b_4 \end{bmatrix},$$

where the following operations result:

$$b_0 = a_0 x_0, \quad b_1 = a_1 x_0 + a_0 x_1, \quad b_2 = a_2 x_0 + a_1 x_1 + a_0 x_2, \quad b_3 = a_3 x_0 + a_2 x_1 + a_1 x_2 + a_0 x_3,$$
$$b_4 = a_4 x_0 + a_3 x_1 + a_2 x_2 + a_1 x_3 + a_0 x_4.$$

As seen in the above matrix, a Toeplitz matrix [T] is a matrix that has constant elements along the main diagonal and the sub-diagonals. Such matrices describe the input-output transformations of the important one-dimensional linear shift invariant (LSI) systems and correlation matrices of stationary sequences [44].

Example 1. Let $u(n) = \{u_0, u_1, u_2, u_3\}$, $w(n) = \{w_0, w_1, w_2, w_3\}$, $\therefore N = 4 \Rightarrow n = 0, 1, \ldots, 2\cdot 4 - 2 = 6$.

$$y(0) = \sum_{k=0}^{3} u(k)w(0-k) = u(0)w(0), \quad y(1) = \sum_{k=0}^{3} u(k)w(1-k) = u(0)w(1) + u(1)w(0),$$

$$y(2) = \sum_{k=0}^{3} u(k)w(2-k) = u(0)w(2) + u(1)w(1) + u(2)w(0),$$

$$y(3) = \sum_{k=0}^{3} u(k)w(3-k) = u(0)w(3) + u(1)w(2) + u(2)w(1) + u(3)w(0),$$

$$y(4) = \sum_{k=0}^{3} u(k)w(4-k) = u(1)w(3) + u(2)w(2) + u(3)w(1),$$

$$y(5) = \sum_{k=0}^{3} u(k)w(5-k) = u(2)w(3) + u(3)w(2), \quad y(6) = \sum_{k=0}^{3} u(k)w(6-k) = u(3)w(3).$$

Figure 3 illustrates the linear convolution implementation using one-dimensional (1D) linear systolic array [42,52-54], where:

$a = u_0 w_1$, $b = u_0 w_2$, $c = u_0 w_3$, $d = u_1 w_1 + u_0 w_2$, $e = u_1 w_2 + u_0 w_3$, $f = u_1 w_3$, $g = u_2 w_1 + u_1 w_2 + u_0 w_3$, $h = u_2 w_2 + u_1 w_3$, $i = u_2 w_3$, $j = u_3 w_1 + u_2 w_2 + u_1 w_3$, $k = u_3 w_2 + u_2 w_3$, $l = u_3 w_3$, $m = u_3 w_2 + u_2 w_3$, $n = u_3 w_3$, $p = u_3 w_3$.

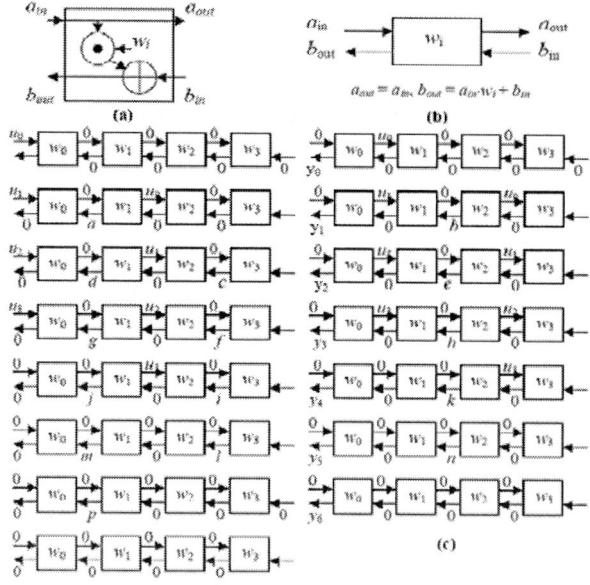

Figure 3. Implementing linear convolution using one-dimensional linear systolic array: (a) cell, (b) block diagram, and (c) left-to-right and top-to-down snapshots of the operation of the systolic array.

2.3. INNER PRODUCT SYSTOLIC ARRAY

Matrix-vector multiplication is fundamental in linear algebraic transformations that model the operation of numerous natural and engineering systems [44,67,68,75]. This type of multiplication can be viewed as the inner product between each matrix row and the transformed vector as follows:

$$\begin{bmatrix} A_{11} & A_{12} & A_{13} & A_{14} \\ A_{21} & A_{22} & A_{23} & A_{24} \\ A_{31} & A_{32} & A_{33} & A_{34} \\ A_{41} & A_{42} & A_{43} & A_{44} \end{bmatrix} \begin{bmatrix} B_1 \\ B_2 \\ B_3 \\ B_4 \end{bmatrix} = \begin{bmatrix} A_{11}B_1 + A_{12}B_2 + A_{13}B_3 + A_{14}B_4 \\ A_{21}B_1 + A_{22}B_2 + A_{23}B_3 + A_{24}B_4 \\ A_{31}B_1 + A_{32}B_2 + A_{33}B_3 + A_{34}B_4 \\ A_{41}B_1 + A_{42}B_2 + A_{43}B_3 + A_{44}B_4 \end{bmatrix}.$$

Figure 4 shows the matrix-vector implementation using the illustrated one-dimensional (1D) systolic array [42,52-54], where:

$a = A_{11}B_1$, $b = A_{21}B_1$, $c = A_{11}B_1 + A_{12}B_2$, $d = A_{31}B_1$, $e = A_{21}B_1 + A_{22}B_2$, $f = A_{11}B_1 + A_{12}B_2 + A_{13}B_3$,
$g = A_{41}B_1$, $h = A_{31}B_1 + A_{32}B_2$, $i = A_{21}B_1 + A_{22}B_2 + A_{23}B_3$, $j = A_{11}B_1 + A_{12}B_2 + A_{13}B_3 + A_{14}B_4$,
$k = A_{41}B_1 + A_{42}B_2$, $l = A_{31}B_1 + A_{32}B_2 + A_{33}B_3$, $m = A_{21}B_1 + A_{22}B_2 + A_{23}B_3 + A_{24}B_4$,
$n = A_{41}B_1 + A_{42}B_2 + A_{43}B_3$, $o = A_{31}B_1 + A_{32}B_2 + A_{33}B_3 + A_{34}B_4$, $p = A_{41}B_1 + A_{42}B_2 + A_{43}B_3 + A_{44}B_4$.

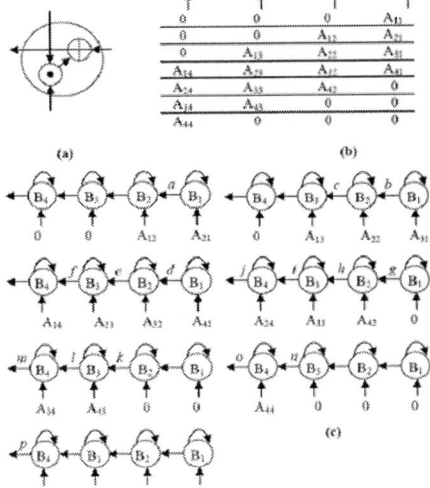

Figure 4. The implementation of inner product utilizing one-dimensional systolic array: (a) cell, (b) architecture, and (c) left-to-right and top-to-down snapshots of the operation of the systolic array.

2.4. BAND MATRIX-VECTOR MULTIPLICATION SYSTOLIC ARRAY

Transforming vectors using band matrices are important in several applications specially that a dense matrix can be considered as a band matrix having the maximum-possible bandwidth [1,10,40,44,66,67,68,75,85]. This type of multiplication can be illustrated as follows:

$$\begin{bmatrix} a_{11} & a_{12} & 0 & 0 & 0 \\ a_{21} & a_{22} & a_{23} & 0 & 0 \\ a_{31} & a_{32} & a_{33} & a_{34} & 0 \\ 0 & a_{42} & a_{43} & a_{44} & a_{45} \\ 0 & 0 & a_{53} & \cdots & \cdots \end{bmatrix} \begin{bmatrix} x_1 \\ x_2 \\ x_3 \\ x_4 \\ x_5 \end{bmatrix} = \begin{bmatrix} y_1 \\ y_2 \\ y_3 \\ y_4 \\ y_5 \end{bmatrix}.$$

where the bandwidth w is computed as $w = 3 + 2 - 1 = 4$. Figure 5 illustrates the band matrix-vector implementation using one-dimensional (1D) systolic array [42,52-54], where the following operations occur:

$y_1 = a_{11}x_1 + a_{12}x_2$, $y_2 = a_{21}x_1 + a_{22}x_2 + a_{23}x_3$, $y_3 = a_{31}x_1 + a_{32}x_2 + a_{33}x_3 + a_{34}x_4$, $y_4 = a_{42}x_2 + a_{43}x_3 + a_{44}x_4 + a_{45}x_5$.

It has been shown that the computation of the n components of y are done in $(2 \cdot n + w)$ time units as compared to $O(n \cdot w)$ time needed for a sequential algorithm on a uni-processor computer [5,8,42,50,52].

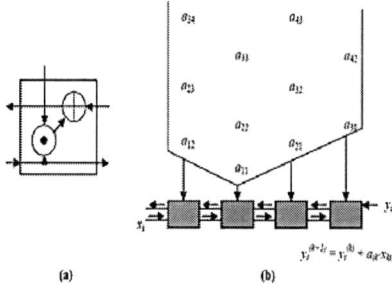

Figure 5. One-dimensional systolic array that implements band matrix – vector multiplication: (a) cell and (b) systolic architecture.

2.5. SORTING SYSTOLIC ARRAY

Sorting is a fundamental operation that exists is several applications [44,53]. Figure 6 shows an example of a systolic sorter that is used to sort the inputs in a descending form [42].

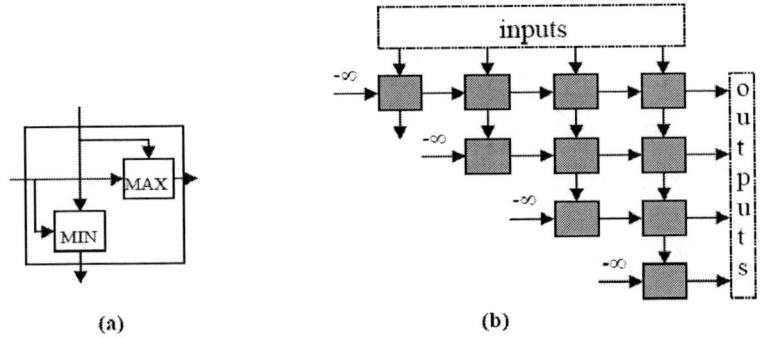

Figure 6. Sorting using a two-dimensional systolic array: (a) cell and (b) systolic network.

Chapter 3

REVERSIBLE CIRCUITS AND SYSTEMS

A (k, k) reversible circuit is a circuit that has the same number of inputs k and outputs k and is a one-to-one mapping between the vectors of inputs and the vectors of outputs, thus the vector of input states can be always uniquely reconstructed from the vector of output states [1,9,18,23,57,66]. Thus, a (k, k) reversible map is a bijective function which is both (1) injective (i.e., "one-to-one" or "(1:1)") and (2) surjective (i.e., "onto"). Such bijective systems are also known as "equipollent", "equipotent", and "one-to-one correspondence". The auxiliary outputs and inputs that are needed only for the purpose of reversibility are called "garbage" outputs and "garbage" inputs respectively. These are auxiliary outputs and inputs from which a reversible map is constructed.

A (k, k) conservative circuit has the same number of inputs k and outputs k and has the same number of values in inputs and outputs (e.g., same number of ones in inputs and outputs for binary, the same number of ones and twos in inputs and outputs for ternary, etc) [1,23]. Figures 7 and 8 show important reversible gates [1,7,66,72]. These gates will be used in later chapters in the reversible implementations of the previously introduced systolic circuits in Sections 2.1 – 2.5, and furthermore the reversible gates in Figure 7 will be used in the quantum realization and computation of the systolic systems in Sections 2.1 – 2.4.

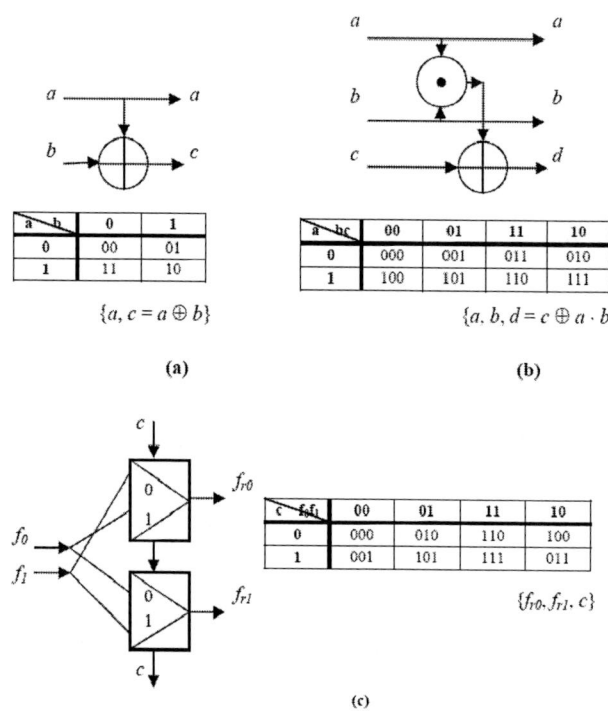

Figure 7. Basic binary reversible primitives and their associated multi-input multi-output K-map representations: (a) (2, 2) Feynman gate, (b) (3, 3) Toffoli gate, and (c) (3, 3) Fredkin gate.

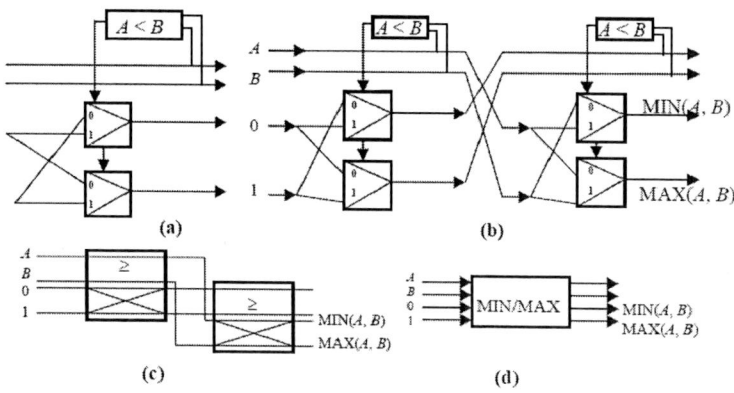

Figure 8. Reversible Min/Max implementation: (a) (4, 4) reversible Picton gate, (b) (4, 4) Min/Max implementation using the Picton gate from (a), (c) schematic circuit for (b), and (d) schematic block diagram for the (4, 4) reversible Min/Max circuit.

Because Galois field (GF) proved to possess desired properties in many applications such as in testing [77], communications, VLSI and signal processing [1,10,12,17,19,40,63,66-68,75,85], the developments of the new systolic logic circuits, in this work, will be conducted on the corresponding Galois field algebraic structures. Radix two and radix three Galois fields (GF(2) and GF(3)) addition and multiplication Tables are defined as shown in Figures 9a and 9b, respectively [1,85].

+	0	1		*	0	1
0	0	1		0	0	0
1	1	0		1	0	1

(a)

+	0	1	2		*	0	1	2
0	0	1	2		0	0	0	0
1	1	2	0		1	0	1	2
2	2	0	1		2	0	2	1

(b)

Figure 9. Tables of Galois field addition and multiplication: (a) radix two and (b) radix three.

The 1-Reduced Post Literal (1-RPL) is a single-variable function which is defined as [1,85]:

$$^i x = \begin{cases} 1, x = i \\ 0, x \neq i \end{cases}. \quad (1)$$

For example, ^0x, ^1x, ^2x are the zero, first, and second polarities of the 1-Reduced Post Literal, respectively.

Also, let us define the ternary shifts of variable x as $\{x, x', x''\}$ to be the zero, first, and second shifts of the variable x, respectively (i.e., $x = x + 0$, $x' = x + 1$, and $x'' = x + 2$), where the utilized variable x can take any value from the set $\{0, 1, 2\}$.

The logic Shannon expansion over third radix Galois field (GF(3)) for a ternary function with a single variable is [1,85]:

$$f = {}^0x f_0 + {}^1x f_1 + {}^2x f_2, \quad (2)$$

where f_0 is the cofactor of f with respect to the variable x of value 0, f_1 is the cofactor of f with respect to the variable x of value 1, and f_2 is the cofactor of f with respect to the variable x of value 2.

Using the addition and multiplication over GF(3), and the axioms of GF(3), it can be shown that the 1-Reduced Post Literals defined in Equation

(1), are related to the shifts of variables over GF(3) in terms of powers as follows [1]:

$$^0x = 2(x)^2 + 1, \tag{3}$$

$$^0x = 2(x')^2 + 2(x'), \tag{4}$$

$$^0x = 2(x'')^2 + x'', \tag{5}$$

$$^1x = 2(x)^2 + 2(x), \tag{6}$$

$$^1x = 2(x')^2 + x', \tag{7}$$

$$^1x = 2(x'')^2 + 1, \tag{8}$$

$$^2x = 2(x)^2 + x, \tag{9}$$

$$^2x = 2(x')^2 + 1, \tag{10}$$

$$^2x = 2(x'')^2 + 2(x''). \tag{11}$$

After the substitution of Equations (3) through (11) in Equation (2), and after the minimization of the terms according to the axioms of the Galois field, one obtains the following Equations:

$$f = 1 \cdot f_0 + x \cdot (2f_1 + f_2) + 2(x)^2 (f_0 + f_1 + f_2), \tag{12}$$

$$f = 1 \cdot f_2 + x' \cdot (2f_0 + f_1) + 2(x')^2 (f_0 + f_1 + f_2), \tag{13}$$

$$f = 1 \cdot f_1 + x'' \cdot (2f_2 + f_0) + 2(x'')^2 (f_0 + f_1 + f_2). \tag{14}$$

Equations (2) and (12) - (14) are the ternary fundamental Shannon and Davio expansions for a single variable, respectively. These equations can be re-written in the following matrix-based forms [1,85]:

$$f = [^0x \; ^1x \; ^2x] \begin{bmatrix} 1 & 0 & 0 \\ 0 & 1 & 0 \\ 0 & 0 & 1 \end{bmatrix} \begin{bmatrix} f_0 \\ f_1 \\ f_2 \end{bmatrix}, \tag{15}$$

$$f = [1 \ x \ x^2] \begin{bmatrix} 1 & 0 & 0 \\ 0 & 2 & 1 \\ 2 & 2 & 2 \end{bmatrix} \begin{bmatrix} f_0 \\ f_1 \\ f_2 \end{bmatrix}, \tag{16}$$

$$f = [1 \ x' \ (x')^2] \begin{bmatrix} 0 & 0 & 1 \\ 2 & 1 & 0 \\ 2 & 2 & 2 \end{bmatrix} \begin{bmatrix} f_0 \\ f_1 \\ f_2 \end{bmatrix}, \tag{17}$$

$$f = [1 \ x'' \ (x'')^2] \begin{bmatrix} 0 & 1 & 0 \\ 1 & 0 & 2 \\ 2 & 2 & 2 \end{bmatrix} \begin{bmatrix} f_0 \\ f_1 \\ f_2 \end{bmatrix}. \tag{18}$$

The extension of the many-valued Shannon and Davio decompositions for two or more variables is obtained using the Kronecker product of the vector of the basis functions and the Kronecker product of the transform matrix [1,85]. To introduce the methodology for the creation of the reversible many-valued Shannon and Davio expansions, the following definitions are introduced [1].

Definition 1. The matrix that is constructed from the permutations of many basis functions of the same type of the corresponding spectral transform is called Generalized Basis Functions Matrix (GBFM).

Definition 2. From the total space of all possible Generalized Basis Function Matrices, the matrices that produce reversible expansions are called Reversible Generalized Basis Functions Matrices.

Example 2. The following is the ternary Shannon transform over GF(3).

$$f = \begin{bmatrix} {}^0c & {}^1c & {}^2c \end{bmatrix} \begin{bmatrix} 1 & 0 & 0 \\ 0 & 1 & 0 \\ 0 & 0 & 1 \end{bmatrix} \begin{bmatrix} f0 \\ f1 \\ f2 \end{bmatrix}.$$

The following is one possible GBFM $\begin{bmatrix} {}^0c & {}^1c & {}^2c \\ {}^0c & {}^2c & {}^1c \\ {}^2c & {}^1c & {}^0c \end{bmatrix}$. Yet, as will be demonstrated in the following method, the upper GBFM is not reversible; it does not produce a reversible expansion. One possible Reversible Shannon

Generalized Basis Function Matrix (RSGBFM) that leads to a reversible expansion is as follows $\begin{bmatrix} {}^0c & {}^1c & {}^2c \\ {}^1c & {}^2c & {}^0c \\ {}^2c & {}^0c & {}^1c \end{bmatrix}$. Note that the Shannon set of basis functions $\{{}^0c, {}^1c, {}^2c\}$ will always appear in any of the rows of the corresponding RSGBFM that produces accordingly the reversible Shannon expansion (in this case, the Shannon set of basis functions $\{{}^0c, {}^1c, {}^2c\}$ appears in the first row).

A necessary and sufficient condition to generate the reversible many-valued Shannon expansions is that the order of the permuted basis functions in the GBFM should satisfy the following constraint [1]: in any given row or column the elements in that row or column are permuted differently than the elements in the corresponding positions of the other rows or columns. This reversibility constraint can be illustrated by means of tables as follows: since the Shannon matrix is orthogonal, then the following (3, 3) ternary Shannon expansion in Equation (19a):

$$\vec{f} = [RSGBFM]\vec{f}_{in} = \begin{bmatrix} {}^2c & {}^0c & {}^1c \\ {}^1c & {}^2c & {}^0c \\ {}^0c & {}^1c & {}^2c \end{bmatrix} \begin{bmatrix} 1 & 0 & 0 \\ 0 & 1 & 0 \\ 0 & 0 & 1 \end{bmatrix} \begin{bmatrix} f_0 \\ f_1 \\ f_2 \end{bmatrix} = \begin{bmatrix} f_{r0} \\ f_{r1} \\ f_{r2} \end{bmatrix}, (19a)$$

is reversible given that input c is produced in the output for which a modified form of Equation (19a) would be as follows:

$$\vec{f} = \begin{bmatrix} {}^2c & {}^0c & {}^1c & 0 \\ {}^1c & {}^2c & {}^0c & 0 \\ {}^0c & {}^1c & {}^2c & 0 \\ 0 & 0 & 0 & 1 \end{bmatrix} \begin{bmatrix} 1 & 0 & 0 & 0 \\ 0 & 1 & 0 & 0 \\ 0 & 0 & 1 & 0 \\ 0 & 0 & 0 & 1 \end{bmatrix} \begin{bmatrix} f_0 \\ f_1 \\ f_2 \\ c \end{bmatrix} = \begin{bmatrix} f_{r0} \\ f_{r1} \\ f_{r2} \\ c \end{bmatrix}. \quad (19b)$$

As an example we are using in this case Shannon set of basis functions $\{{}^0c, {}^1c, {}^2c\}$ that appear in the third row, but Shannon set of basis functions $\{{}^0c, {}^1c, {}^2c\}$ can appear in any of the rows of the corresponding RGBFM. The ternary Shannon expansion in Equation (19) is reversible as shown in Table 1 using the permutation of cofactors [1].

Table 1. Reversibility proof using the permutation of cofactors for the ternary Shannon expansion in Eq. (19).

Function \ 1-RPL	^0c	^1c	^2c
$f =$	f_1	f_2	f_0
$f =$	f_2	f_0	f_1
$f =$	f_0	f_1	f_2

Thus one can observe that, due to the uniqueness in the 1-RPL selection, the resulting function values (i.e., cofactors f_0, f_1, and f_2) for different 1-RPL values are always distinct (i.e., mutually exclusive). The same method can be carried out for any configuration that reflects RSGBFM for any radix.

Example 3. The following are all possible permutations of the Reversible Generalized Basis Functions Matrix for the Shannon transform matrix $\begin{bmatrix} 1 & 0 & 0 \\ 0 & 1 & 0 \\ 0 & 0 & 1 \end{bmatrix}$ to produce the corresponding reversible ternary Shannon expansions over GF(3) [1,3] given that input c is produced in the output.

$$\vec{f} = \begin{bmatrix} ^0c & ^1c & ^2c \\ ^1c & ^2c & ^0c \\ ^2c & ^0c & ^1c \end{bmatrix} \begin{bmatrix} 1 & 0 & 0 \\ 0 & 1 & 0 \\ 0 & 0 & 1 \end{bmatrix} \begin{bmatrix} f0 \\ f1 \\ f2 \end{bmatrix} = \begin{bmatrix} fr0 \\ fr1 \\ fr2 \end{bmatrix}, \quad (20)$$

$$\vec{f} = \begin{bmatrix} ^0c & ^1c & ^2c \\ ^2c & ^0c & ^1c \\ ^1c & ^2c & ^0c \end{bmatrix} \begin{bmatrix} 1 & 0 & 0 \\ 0 & 1 & 0 \\ 0 & 0 & 1 \end{bmatrix} \begin{bmatrix} f0 \\ f1 \\ f2 \end{bmatrix} = \begin{bmatrix} fr0 \\ fr1 \\ fr2 \end{bmatrix}, \quad (21)$$

$$\vec{f} = \begin{bmatrix} ^1c & ^2c & ^0c \\ ^0c & ^1c & ^2c \\ ^2c & ^0c & ^1c \end{bmatrix} \begin{bmatrix} 1 & 0 & 0 \\ 0 & 1 & 0 \\ 0 & 0 & 1 \end{bmatrix} \begin{bmatrix} f0 \\ f1 \\ f2 \end{bmatrix} = \begin{bmatrix} fr0 \\ fr1 \\ fr2 \end{bmatrix}, \quad (22)$$

$$\vec{f} = \begin{bmatrix} ^2c & ^0c & ^1c \\ ^0c & ^1c & ^2c \\ ^1c & ^2c & ^0c \end{bmatrix} \begin{bmatrix} 1 & 0 & 0 \\ 0 & 1 & 0 \\ 0 & 0 & 1 \end{bmatrix} \begin{bmatrix} f0 \\ f1 \\ f2 \end{bmatrix} = \begin{bmatrix} fr0 \\ fr1 \\ fr2 \end{bmatrix}, \quad (23)$$

$$\vec{f} = \begin{bmatrix} ^1c & ^2c & ^0c \\ ^2c & ^0c & ^1c \\ ^0c & ^1c & ^2c \end{bmatrix} \begin{bmatrix} 1 & 0 & 0 \\ 0 & 1 & 0 \\ 0 & 0 & 1 \end{bmatrix} \begin{bmatrix} f0 \\ f1 \\ f2 \end{bmatrix} = \begin{bmatrix} fr0 \\ fr1 \\ fr2 \end{bmatrix}, \quad (24)$$

$$\vec{f} = \begin{bmatrix} {}^2c & {}^0c & {}^1c \\ {}^1c & {}^2c & {}^0c \\ {}^0c & {}^1c & {}^2c \end{bmatrix} \begin{bmatrix} 1 & 0 & 0 \\ 0 & 1 & 0 \\ 0 & 0 & 1 \end{bmatrix} \begin{bmatrix} f_0 \\ f_1 \\ f_2 \end{bmatrix} = \begin{bmatrix} fr_0 \\ fr_1 \\ fr_2 \end{bmatrix}. \tag{25}$$

In the special case of binary XOR (i.e., GF(2)) logic, there are only two reversible Shannon gates as follows [1,2]:

$$\vec{f} = \begin{bmatrix} \bar{c} & c \\ c & \bar{c} \end{bmatrix} \begin{bmatrix} 1 & 0 \\ 0 & 1 \end{bmatrix} \begin{bmatrix} f_0 \\ f_1 \end{bmatrix} = \begin{bmatrix} fr_0 \\ fr_1 \end{bmatrix}, \tag{26}$$

$$\vec{f} = \begin{bmatrix} c & \bar{c} \\ \bar{c} & c \end{bmatrix} \begin{bmatrix} 1 & 0 \\ 0 & 1 \end{bmatrix} \begin{bmatrix} f_0 \\ f_1 \end{bmatrix} = \begin{bmatrix} fr_0 \\ fr_1 \end{bmatrix}. \tag{27}$$

Reversible many-valued Davio expansions have been also created [1,3]. For example, for the following reversible ternary Shannon expansion:

$$\vec{f} = \begin{bmatrix} {}^0c & {}^1c & {}^2c \\ {}^1c & {}^2c & {}^0c \\ {}^2c & {}^0c & {}^1c \end{bmatrix} \begin{bmatrix} 1 & 0 & 0 \\ 0 & 1 & 0 \\ 0 & 0 & 1 \end{bmatrix} \begin{bmatrix} f_0 \\ f_1 \\ f_2 \end{bmatrix}, \tag{28}$$

there exist three Davio types for each row of the RSGBFM. Utilizing the derivation of the ternary Davio expansion (as in Equations (12) – (14)), the following are the D_0-type expansions for the first row, second row, and third row of the above RSGBFM $\begin{bmatrix} {}^0c & {}^1c & {}^2c \\ {}^1c & {}^2c & {}^0c \\ {}^2c & {}^0c & {}^1c \end{bmatrix}$, respectively:

$$f_{D0,row0} = [1 \ c \ c^2] \begin{bmatrix} 1 & 0 & 0 \\ 0 & 2 & 1 \\ 2 & 2 & 2 \end{bmatrix} \begin{bmatrix} f_0 \\ f_1 \\ f_2 \end{bmatrix}, \tag{29}$$

$$f_{D0,row1} = [1 \ c \ c^2] \begin{bmatrix} 0 & 0 & 1 \\ 2 & 1 & 0 \\ 2 & 2 & 2 \end{bmatrix} \begin{bmatrix} f_0 \\ f_1 \\ f_2 \end{bmatrix}, \tag{30}$$

$$f_{D0,row2} = [1 \ c \ c^2] \begin{bmatrix} 0 & 1 & 0 \\ 1 & 0 & 2 \\ 2 & 2 & 2 \end{bmatrix} \begin{bmatrix} f_0 \\ f_1 \\ f_2 \end{bmatrix}. \tag{31}$$

To produce one form of the reversible Davio$_0$-type function expansion, let us choose the transform matrix $\begin{bmatrix} 0 & 0 & 1 \\ 2 & 1 & 0 \\ 2 & 2 & 2 \end{bmatrix}$ in Equation (30) to produce the corresponding reversible many-valued Davio$_0$ expansion. Note that we have two other choices of $\begin{bmatrix} 1 & 0 & 0 \\ 0 & 2 & 1 \\ 2 & 2 & 2 \end{bmatrix}$ and $\begin{bmatrix} 0 & 1 & 0 \\ 1 & 0 & 2 \\ 2 & 2 & 2 \end{bmatrix}$ transform matrices in Equations (29) and (31), respectively. It has been shown [1,3] that for the above transform matrix the following is the ternary reversible Davio$_0$ expansion:

$$\vec{f}_{D0} = \begin{bmatrix} 1 & 1+c & 1+2c+c^2 \\ 1 & c & c^2 \\ 1 & 2+c & 1+c+c^2 \end{bmatrix} \begin{bmatrix} 0 & 0 & 1 \\ 2 & 1 & 0 \\ 2 & 2 & 2 \end{bmatrix} \begin{bmatrix} f_0 \\ f_1 \\ f_2 \end{bmatrix} = \begin{bmatrix} 1 & 1+c & 1+2c+c^2 \\ 1 & c & c^2 \\ 1 & 2+c & 1+c+c^2 \end{bmatrix} \begin{bmatrix} f_2 \\ 2f_0+f_1 \\ 2f_0+2f_1+2f_2 \end{bmatrix} \quad (32\text{a,b,c})$$

$$= \begin{bmatrix} f_2 + (1+c)(2f_0+f_1) + (1+2c+c^2)(2f_0+2f_1+2f_2) \\ f_2 + c(2f_0+f_1) + c^2(2f_0+2f_1+2f_2) \\ f_2 + (2+c)(2f_0+f_1) + (1+c+c^2)(2f_0+2f_1+2f_2) \end{bmatrix}$$

$$= \begin{bmatrix} f_0 + c(2f_1+f_2) + c^2(2f_0+2f_1+2f_2) \\ f_2 + c(2f_0+f_1) + c^2(2f_0+2f_1+2f_2) \\ f_1 + c(2f_2+f_0) + c^2(2f_0+2f_1+2f_2) \end{bmatrix} = \begin{bmatrix} f_{r0,D0} \\ f_{r1,D0} \\ f_{r2,D0} \end{bmatrix}$$

Figure 10 shows important ternary radix GF-based reversible gates [1]. These gates will be used in later sections in the ternary reversible implementation, and the extension to ternary quantum implementation, of the classical ternary systolic arrays.

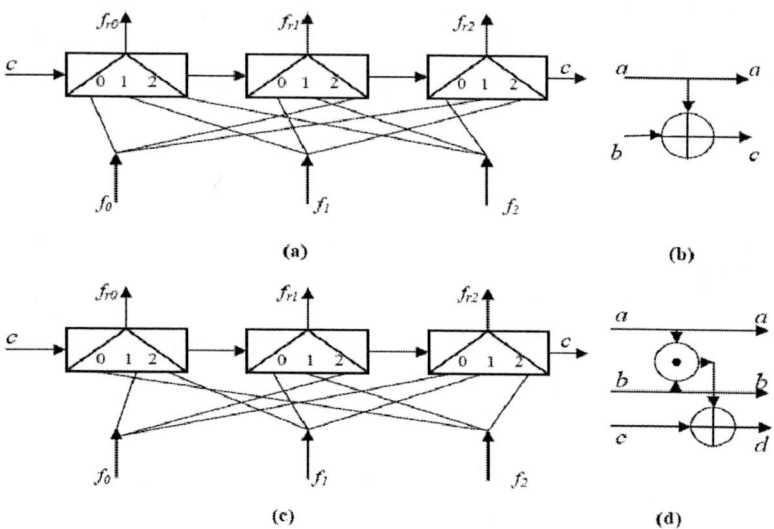

Figure 10. Basic ternary reversible primitives: (a) (4, 4) Shannon (also called Fredkin) gate for Equation (21), (b) (2, 2) Feynman gate, (c) (4, 4) Shannon (also called Fredkin) gate for Equation (24), and (d) (3, 3) Toffoli gate.

For reversibility illustration, Table 2 presents reversibility proof for the ternary Toffoli gate where the inputs are in order $\{a, b, c\}$ and outputs are in order $\{a, b, d = c +_3 a *_3 b\}$.

Table 2. Proving reversibility for the ternary (3, 3) Toffoli gate where the inputs are in order $\{a, b, c\}$ and outputs are in order $\{a, b, d = c +_3 a *_3 b\}$.

000	000	100	100	200	200
001	001	101	101	201	201
002	002	102	102	202	202
010	010	110	111	210	212
011	011	111	112	211	210
012	012	112	110	212	211
020	020	120	122	220	221
021	021	121	120	221	222
022	022	122	121	222	220

Chapter 4

REVERSIBLE SYSTOLIC NETWORKS

This chapter introduces new results for the synthesis of m-ary Galois functions using reversible systolic networks using the following two methods: (1) reusing the classical add-multiply PE cells by showing their m-ary Galois logic reversibility and thereafter reusing the whole systolic structure since the connection of reversible PEs will produce by necessity a reversible systolic circuit (cf. Figure 11), (since the interconnection of reversible gates will necessarily produce a reversible circuit) [1], and (2) direct mapping of classical irreversible systolic circuits into reversible systolic circuits by interconnecting the reversible counterpart of the irreversible PEs (cf. Figure 15). Moreover, the implementation of the m-ary functions using the new reversible systolic arrays is performed through two methods (cf. Example 4): (a) direct implementation by the substitution of the array's values using the corresponding m-ary functional literals, and (b) implementation using the corresponding m-ary matrix-based elements that are obtained through the m-ary reversible Shannon and Davio functional expansions.

Basic PEs which are used in the construction of arithmetic systolic arrays are the add-multiply cells [42,52-54]. Figure 11a introduces the binary (two-valued) (3, 3) reversible Toffoli gate (cf. Figure 7b) that implements reversibly (in the second radix Galois logic GF(2)) the classical Kung add-multiply cell (in Figure 11b). Since the interconnection of reversible PEs will produce by necessity a reversible systolic circuit, Figure 11c implements the 2D reversible Kung systolic array over GF(2) by interconnecting the GF(2) (3, 3) reversible Toffoli gates.

The implementation using the corresponding GF(2) matrix-based elements that can be obtained through the GF(2) reversible Shannon functional expansions from Equation (26) is obtained as follows:

$$\begin{bmatrix} 0 & 0 & 0 & 0 & 0 \\ 0 & {}^0e & {}^1e & 0 & 0 \\ 0 & {}^1e & {}^0e & 0 & 0 \\ 0 & 0 & 0 & 0 & 0 \\ 0 & 0 & 0 & 0 & 0 \end{bmatrix} \cdot \begin{bmatrix} 0 & 0 & 0 & 0 & 0 \\ 0 & f_0 & 0 & 0 & 0 \\ 0 & f_1 & 0 & 0 & 0 \\ 0 & 0 & 0 & 0 & 0 \\ 0 & 0 & 0 & 0 & 0 \end{bmatrix} = \begin{bmatrix} 0 & 0 & 0 & 0 & 0 \\ 0 & {}^0ef_0 + {}^1ef_1 & 0 & 0 & 0 \\ 0 & {}^1ef_0 + {}^0ef_1 & 0 & 0 & 0 \\ 0 & 0 & 0 & 0 & 0 \\ 0 & 0 & 0 & 0 & 0 \end{bmatrix} \qquad (33)$$

where:

$a_{22} = {}^0e, a_{23} = {}^1e, a_{32} = {}^1e, a_{33} = {}^0e, b_{22} = f_0, b_{32} = f_1, c_{22} = {}^0ef_0 + {}^1ef_1, c_{32} = {}^1ef_0 + {}^0ef_1$.

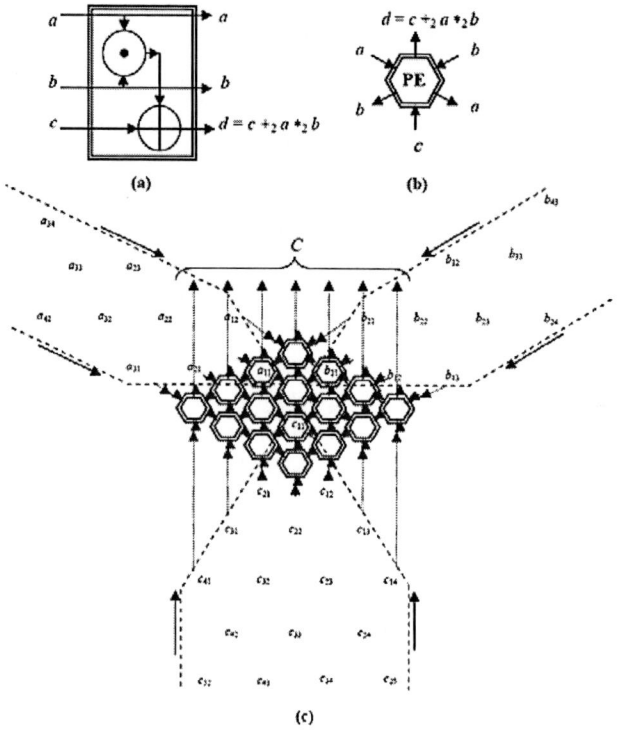

Figure 11. Two-valued Galois reversible Kung systolic network: (a) reversible (3, 3) Toffoli gate, (b) reversible (3, 3) Kung cell, and (c) reversible Kung systolic network.

For the systolic implementation of m-ary functions using the corresponding GF(3), as an example of an m-ary GF logic, the matrix-based elements that are obtained using the GF(3) reversible Shannon and Davio functional expansions from Equations (20) and (32) are obtained as follows, respectively:

$$\begin{bmatrix} 0 & 0 & 0 & 0 & 0 \\ ^0e & ^1e & ^2e & 0 & 0 \\ ^1e & ^2e & ^0e & 0 & 0 \\ 0 & ^0e & ^1e & ^2e & 0 \\ 0 & 0 & 0 & 0 & 0 \end{bmatrix} \cdot \begin{bmatrix} 0 & 0 & f_0 & 0 & 0 \\ 0 & 0 & f_1 & 0 & 0 \\ 0 & 0 & f_2 & 0 & 0 \\ 0 & 0 & f_0 & 0 & 0 \\ 0 & 0 & 0 & 0 & 0 \end{bmatrix} = \begin{bmatrix} 0 & 0 & 0 & 0 & 0 \\ 0 & 0 & ^0ef_0+^1ef_1+^2ef_2 & 0 & 0 \\ 0 & 0 & ^1ef_0+^2ef_1+^0ef_2 & 0 & 0 \\ 0 & 0 & ^0ef_1+^1ef_2+^2ef_0 & 0 & 0 \\ 0 & 0 & 0 & 0 & 0 \end{bmatrix} \quad (34)$$

where:

$a_{21}=^0e, a_{22}=^1e, a_{23}=^2e, a_{31}=^1e, a_{32}=^2e, a_{33}=^0e, a_{42}=^0e, a_{43}=^1e, a_{44}=^2e, b_{13}=f_0, b_{23}=f_1, b_{33}=f_2, b_{43}=f_0$, $c_{23}=^0ef_0+^1ef_1+^2ef_2, c_{33}=^1ef_0+^2ef_1+^0ef_2, c_{43}=^2ef_0+^0ef_1+^1ef_2$.

$$\begin{bmatrix} 0 & 0 & 0 & 0 & 0 \\ 1 & 1+e & 1+2e+e^2 & 0 & 0 \\ 1 & e & e^2 & 0 & 0 \\ 0 & 2+e & 1+e+e^2 & 1 & 0 \\ 0 & 0 & 0 & 0 & 0 \end{bmatrix} \cdot \begin{bmatrix} 0 & 0 & f_2 & 0 & 0 \\ 0 & 0 & 2f_0+f_1 & 0 & 0 \\ 0 & 0 & 2f_0+2f_1+2f_2 & 0 & 0 \\ 0 & 0 & f_2 & 0 & 0 \\ 0 & 0 & 0 & 0 & 0 \end{bmatrix} =$$

$$\begin{bmatrix} 0 & 0 & 0 & 0 & 0 \\ 0 & 0 & f_2+(1+e)(2f_0+f_1)+(1+2e+e^2)(2f_0+2f_1+2f_2) & 0 & 0 \\ 0 & 0 & f_2+e(2f_0+f_1)+e^2(2f_0+2f_1+2f_2) & 0 & 0 \\ 0 & 0 & f_2+(2+e)(2f_0+f_1)+(1+e+e^2)(2f_0+2f_1+2f_2) & 0 & 0 \\ 0 & 0 & 0 & 0 & 0 \end{bmatrix} \quad (35)$$

where:

$a_{21}=1, a_{22}=1+e, a_{23}=1+e+e^2, a_{31}=1, a_{32}=e, a_{33}=e^2, a_{42}=2+e, a_{43}=1+e+e^2$, $a_{44}=1, b_{13}=f_2, b_{23}=2f_0+f_1, b_{33}=(2f_0+2f_1+2f_2), b_{43}=f_2$,
$c_{23}=f_2+(1+e)(2f_0+f_1)+(1+2e+e^2)(2f_0+2f_1+2f_2)$,
$c_{33}=f_2+e(2f_0+f_1)+e^2(2f_0+2f_1+2f_2), c_{43}=f_2+(2+e)(2f_0+f_1)+(1+e+e^2)(2f_0+2f_1+2f_2)$.

The realization of a many-variable function using the methods shown in Equations (33) – (35) follow the matrix-based expansion using the Kronecker (i.e., tensor) product [1] and the use of Toffoli PE (TPE) over GF(3) (cf. Figure 10d and Table 2).

The temporal complexity of the introduced reversible systolic array in Figure 11 can be performed as the new m-ary reversible systolic array can finish the band matrix multiplication in T time units, where: $T = 3n + \min(w_1, w_2)$. Therefore the computation time is linearly proportional to the dimension n of the matrix, and when the matrix bandwidths increase to $w_1 = w_2 = n$ (for dense matrices [**A**] and [**B**]), the time becomes $O(4n)$, neglecting the I/O time delays. If one used a single non-systolic additive-multiply processor (i.e., circuit) to perform the same matrix multiplication, $O(n^3)$ computation time would be needed. The new m-ary reversible systolic array thus has a speed gain of $O(n^2)$, and this becomes more apparent for large n.

The spatial complexity (i.e., garbage count) of the introduced reversible systolic array in Figure 11 can be performed as the new m-ary reversible systolic array has garbage outputs (i.e., auxiliary outputs that are needed only for the purpose of reversibility) at the bottom-left and bottom-right equal to $2\sqrt{w1 \cdot w2}$. Thus, for the reversible systolic array in Figure 11, one has 8 garbage outputs.

As technology mapping is important in several digital design applications [13,64], the following algorithm, which is called the Reversible Gate Mapping (RGM) algorithm, shows an XOR-based (i.e., Boolean difference) method to map between various reversible gates over GF(2), where XNOR (which is the Boolean equivalence and also is the Boolean compliment of XOR) can be used similarly in the RGM algorithm as well.

Algorithm RGM

1. $\text{Map}_2 = \text{Map}_1 \oplus \text{Map}_{correction}$, $\text{Map}_1 = \text{Map}_2 \oplus \text{Map}_{correction}$.
2. Using Logic Optimization technique, obtain SOP/POS optimized functions' forms.
3. Synthesize the solution using the Boolean difference (XOR) between the corresponding maps.

The RGM algorithm can be used to replace reversible gates with other kind of reversible gates that are more suitable for synthesis, manufacturing, or for different types of applications. As will be shown in Chapter 5, that the reversible Controlled-Not-based and Controlled-Swap-based gates are essential in the quantum circuit synthesis, Figure 12 illustrates an example of the implementation of the RGM algorithm for the mapping and inverse mapping between (3, 3) Toffoli and (3, 3) Fredkin reversible gates, and Figure 13 shows the resulting circuits for such mapping, where it is shown in Figures 13a and 13b that three Toffoli gates plus five Feynman gates produces one Fredkin gate, and is shown in Figures 13a and 13c that two Toffoli gates plus five Feynman gates and one Fredkin gate will produce one Toffoli gate.

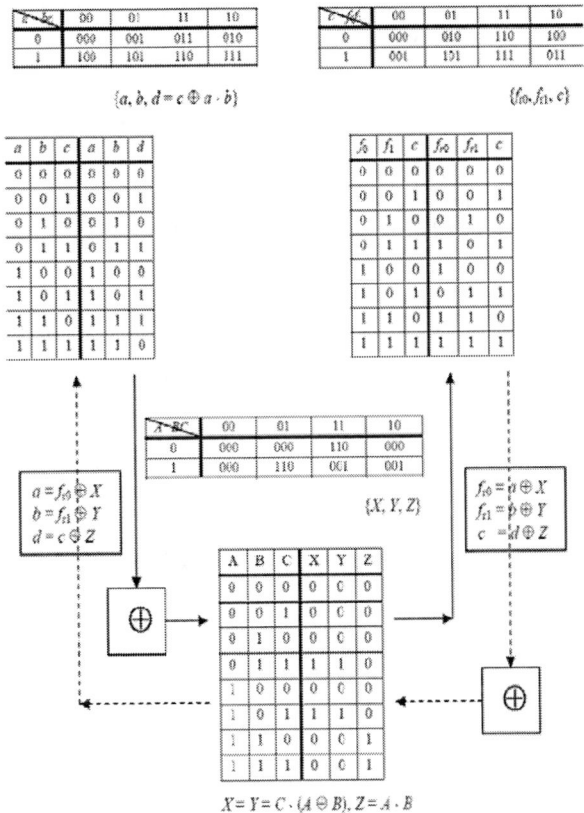

Figure 12. The implementation of the Reversible Gate Mapping (RGM) algorithm for the mapping and inverse mapping between (3, 3) Toffoli and Fredkin reversible primitives.

Multiple-level reversible circuits are also possible by the iterative use (i.e., looping) of the RGM algorithm on the correction part (i.e., serial-mode

RGM) or on both the correction part and the error part (i.e., the first part) at the same time (i.e., parallel-mode RGM). Also, the iterative use of the RGM method using a hybrid of both XOR ($\oplus$) (i.e., Boolean difference) and XNOR ($\otimes$) (i.e., Boolean equivalence) operations to decompose Boolean functions can also be accomplished [4].

Although the results that are presented in this chapter are illustrated for the case of the 2D hexagonal systolic array, and since the basic PE in these circuits is the add-multiply cell and Toffoli cell (TPE) is the reversible GF counterpart of this fundamental add-multiply PE, the systolic arrays in Sections 2.2 – 2.4 (cf. Figures 3 – 5) can be implemented reversibly using the interconnection between TPEs as well.

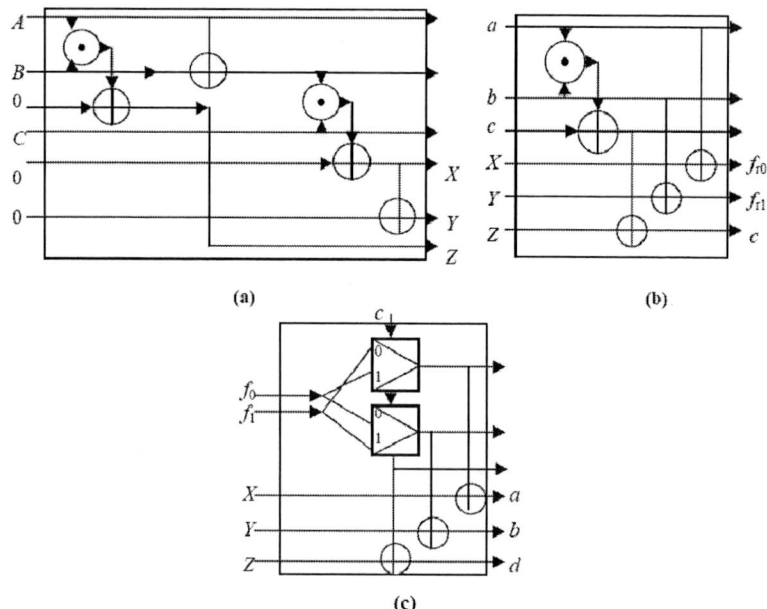

Figure 13. Using the RGM algorithm for the mapping between reversible primitives: (a) generated mapping circuit, (b) three Toffoli gates plus five Feynman gates that produce one Fredkin gate, and (c) two Toffoli gates plus five Feynman gates and one Fredkin gate that produce one Toffoli gate.

The RGM algorithm can be generalized to any many-valued GF radix, where the corresponding many-valued GF addition and multiplication operators and many-valued reversible gates (e.g., Figure 10) must be used in the mapping.

Example 4 shows the bijective implementation, using the reversible Kung systolic network from Figure 11, of the two-digit ternary multiplication which is performed utilizing the mod-multiplication operator.

Example 4. Figure 14 shows the maps for the ternary multiplication (M) and the output carry (C_{out}).

a\b	0	1	2		a\b	0	1	2
0	0	0	0		0	0	0	0
1	0	1	2		1	0	0	0
2	0	2	1		2	0	0	1

              a   b

Figure 14. Ternary two-digit multiplier using ternary mod-multiplication: (a) ternary multiplication (M) and (b) ternary carry out (C_{out}).

The following is a bijective implementation, using the GF(2) reversible Kung systolic network from Figure 11, of the two-digit ternary multiplication which is performed utilizing the mod-multiplication operator.

$M = {}^1a^1b + 2 \cdot {}^2a^1b + 2 \cdot {}^1a^2b + {}^2a^2b,$

$C_{out} = {}^2a^2b,$

$\rightarrow c_{33} = M, a_{31} = {}^1a, b_{13} = {}^1b, a_{32} = 2 \cdot {}^2a, b_{23} = {}^1b, a_{33} = 2 \cdot {}^1a, b_{33} = {}^2b, a_{34} = {}^2a, b_{43} = {}^2b.$

$\rightarrow c_{11} = C_{out}, a_{11} = {}^2a, b_{11} = {}^2b, a_{12} = 0, b_{21} = 0.$

This upper implementation follows directly by the substitution of the array's values using the corresponding m-ary functional literals. A second way to implement the upper multiplier is by using the many-variable ternary Kronecker-based reversible Shannon and Davio expansions of Equation (20) and (32) that are obtained as follows, respectively, where the values of the arrays [A] and [B] (in Figure 11) are given the corresponding values from the following reversible Shannon and Davio expansions in a manner similar to that used in Equations (33) – (35).

$$\vec{f} = \begin{bmatrix} f_{r0} \\ f_{r1} \\ f_{r2} \\ f_{r3} \\ f_{r4} \\ f_{r5} \\ f_{r6} \\ f_{r7} \\ f_{r8} \end{bmatrix} = \left(\begin{bmatrix} 0_a & 1_a & 2_a \\ 1_a & 2_a & 0_a \\ 2_a & 0_a & 1_a \end{bmatrix} \otimes \begin{bmatrix} 0_b & 1_b & 2_b \\ 1_b & 2_b & 0_b \\ 2_b & 0_b & 1_b \end{bmatrix} \right) \left(\begin{bmatrix} 1 & 0 & 0 \\ 0 & 1 & 0 \\ 0 & 0 & 1 \end{bmatrix} \otimes \begin{bmatrix} 1 & 0 & 0 \\ 0 & 1 & 0 \\ 0 & 0 & 1 \end{bmatrix} \right) \left(\begin{bmatrix} f_{a0} \\ f_{a1} \\ f_{a2} \end{bmatrix} \otimes \begin{bmatrix} f_{b0} \\ f_{b1} \\ f_{b2} \end{bmatrix} \right)$$

$$= \begin{bmatrix} 0_a0_b & 0_a1_b & 0_a2_b & 1_a0_b & 1_a1_b & 1_a2_b & 2_a0_b & 2_a1_b & 2_a2_b \\ 0_a1_b & 0_a2_b & 0_a0_b & 1_a1_b & 1_a2_b & 1_a0_b & 2_a1_b & 2_a2_b & 2_a0_b \\ 0_a2_b & 0_a0_b & 0_a1_b & 1_a2_b & 1_a0_b & 1_a1_b & 2_a2_b & 2_a0_b & 2_a1_b \\ 1_a0_b & 1_a1_b & 1_a2_b & 2_a0_b & 2_a1_b & 2_a2_b & 0_a0_b & 0_a1_b & 0_a2_b \\ 1_a1_b & 1_a2_b & 1_a0_b & 2_a1_b & 2_a2_b & 2_a0_b & 0_a1_b & 0_a2_b & 0_a0_b \\ 1_a2_b & 1_a0_b & 1_a1_b & 2_a2_b & 2_a0_b & 2_a1_b & 0_a2_b & 0_a0_b & 0_a1_b \\ 2_a0_b & 2_a1_b & 2_a2_b & 0_a0_b & 0_a1_b & 0_a2_b & 1_a0_b & 1_a1_b & 1_a2_b \\ 2_a1_b & 2_a2_b & 2_a0_b & 0_a1_b & 0_a2_b & 0_a0_b & 1_a1_b & 1_a2_b & 1_a0_b \\ 2_a2_b & 2_a0_b & 2_a1_b & 0_a2_b & 0_a0_b & 0_a1_b & 1_a2_b & 1_a0_b & 1_a1_b \end{bmatrix} \begin{bmatrix} 1 & 0 & 0 & 0 & 0 & 0 & 0 & 0 & 0 \\ 0 & 1 & 0 & 0 & 0 & 0 & 0 & 0 & 0 \\ 0 & 0 & 1 & 0 & 0 & 0 & 0 & 0 & 0 \\ 0 & 0 & 0 & 1 & 0 & 0 & 0 & 0 & 0 \\ 0 & 0 & 0 & 0 & 1 & 0 & 0 & 0 & 0 \\ 0 & 0 & 0 & 0 & 0 & 1 & 0 & 0 & 0 \\ 0 & 0 & 0 & 0 & 0 & 0 & 1 & 0 & 0 \\ 0 & 0 & 0 & 0 & 0 & 0 & 0 & 1 & 0 \\ 0 & 0 & 0 & 0 & 0 & 0 & 0 & 0 & 1 \end{bmatrix} \begin{bmatrix} f_{a0b0} \\ f_{a0b1} \\ f_{a0b2} \\ f_{a1b0} \\ f_{a1b1} \\ f_{a1b2} \\ f_{a2b0} \\ f_{a2b1} \\ f_{a2b2} \end{bmatrix}$$

$$\vec{f} = \begin{bmatrix} f_{r0} \\ f_{r1} \\ f_{r2} \\ f_{r3} \\ f_{r4} \\ f_{r5} \\ f_{r6} \\ f_{r7} \\ f_{r8} \end{bmatrix} = \left(\begin{bmatrix} 1 & 1+a & 1+2a+a^2 \\ 1 & a & a^2 \\ 1 & 2+a & 1+a+a^2 \end{bmatrix} \otimes \begin{bmatrix} 1 & 1+b & 1+2b+b^2 \\ 1 & b & b^2 \\ 1 & 2+b & 1+b+b^2 \end{bmatrix} \right) \left(\begin{bmatrix} 0 & 0 & 1 \\ 2 & 1 & 0 \\ 2 & 2 & 2 \end{bmatrix} \otimes \begin{bmatrix} 0 & 0 & 1 \\ 2 & 1 & 0 \\ 2 & 2 & 2 \end{bmatrix} \right) \left(\begin{bmatrix} f_{a0} \\ f_{a1} \\ f_{a2} \end{bmatrix} \otimes \begin{bmatrix} f_{b0} \\ f_{b1} \\ f_{b2} \end{bmatrix} \right)$$

$$= \begin{bmatrix} 1 & 1+b & 1+2b+b^2 & 1+a & (1+a)(1+b) & (1+a)(1+2b+b^2) & 1+2a+a^2 & (1+2a+a^2)(1+b) & (1+2a+a^2)(1+2b+b^2) \\ 1 & b & b^2 & 1+a & (1+a)b & (1+a)b^2 & 1+2a+a^2 & (1+2a+a^2)b & (1+2a+a^2)b^2 \\ 1 & 2+b & 1+b+b^2 & 1+a & (1+a)(2+b) & (1+a)(1+b+b^2) & 1+2a+a^2 & (1+2a+a^2)(2+b) & (1+2a+a^2)(1+b+b^2) \\ 1 & 1+b & 1+2b+b^2 & a & a(1+b) & a(1+2b+b^2) & a^2 & a^2(1+b) & a^2(1+2b+b^2) \\ 1 & b & b^2 & a & ab & ab^2 & a^2 & a^2b & a^2b^2 \\ 1 & 2+b & 1+b+b^2 & a & a(2+b) & a(1+b+b^2) & a^2 & a^2(2+b) & a^2(1+b+b^2) \\ 1 & 1+b & 1+2b+b^2 & 2+a & (2+a)(1+b) & (2+a)(1+2b+b^2) & 1+a+a^2 & (1+a+a^2)(1+b) & (1+a+a^2)(1+2b+b^2) \\ 1 & b & b^2 & 2+a & (2+a)b & (2+a)b^2 & 1+a+a^2 & (1+a+a^2)b & (1+a+a^2)b^2 \\ 1 & 2+b & 1+b+b^2 & 2+a & (2+a)(2+b) & (2+a)(1+b+b^2) & 1+a+a^2 & (1+a+a^2)(2+b) & (1+a+a^2)(1+b+b^2) \end{bmatrix} \begin{bmatrix} 0 & 0 & 0 & 0 & 0 & 0 & 0 & 0 & 1 \\ 0 & 0 & 0 & 0 & 0 & 0 & 2 & 1 & 0 \\ 0 & 0 & 0 & 0 & 0 & 0 & 2 & 2 & 2 \\ 0 & 0 & 2 & 0 & 0 & 1 & 0 & 0 & 0 \\ 1 & 2 & 0 & 2 & 1 & 0 & 0 & 0 & 0 \\ 1 & 1 & 1 & 2 & 2 & 2 & 0 & 0 & 0 \\ 0 & 0 & 2 & 0 & 0 & 2 & 0 & 0 & 2 \\ 1 & 2 & 0 & 1 & 2 & 0 & 1 & 2 & 0 \\ 1 & 1 & 1 & 1 & 1 & 1 & 1 & 1 & 1 \end{bmatrix} \begin{bmatrix} f_{a0b0} \\ f_{a0b1} \\ f_{a0b2} \\ f_{a1b0} \\ f_{a1b1} \\ f_{a1b2} \\ f_{a2b0} \\ f_{a2b1} \\ f_{a2b2} \end{bmatrix}$$

In Example 4, the implementation of the second method (utilizing Figure 11c) can be done in two ways: (1) using a single matrix-matrix multiplication systolic array that multiplies the first basis vector matrix by the second vector of weighted sum of cofactors, and (2) using two chained matrix-matrix multiplication systolic arrays where the first systolic array multiplies the spectral transform matrix [S] and the cofactors vector and the second systolic array multiplies the basis vector matrix and the output vector that results from the first systolic array.

Although, as shown in Example 4, the reversible systolic implementation method using the reversible expansions is more general and more systematic than the direct implementation method without using reversible expansions, yet the reversible systolic implementation method using the reversible expansions is more expensive than the direct implementation method without using reversible expansions in terms of the number of array elements in arrays [**A**] and [**B**] that are needed to be used.

Another method of producing reversible systolic circuits is by mapping the classical irreversible systolic circuits into reversible systolic circuits by interconnecting the reversible counterpart of the irreversible PEs. As an example, Figure 15 illustrates the reversible Min/Max implementation for the sorting operation using the reversible Min/Max gate from Figure 8, which implements reversibly the Min/Max circuit from Figure 6.

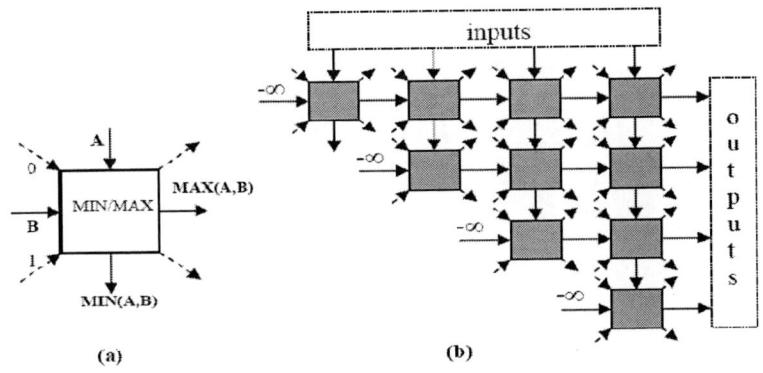

Figure 15. The reversible realization of the Min/Max systolic network: (a) (4, 4) reversible Min/Max cell from Figure 8d and (b) the reversible systolic Min/Max network.

As mentioned previously, although the TPE-based results that are presented in this chapter are illustrated for the case of the 2D hexagonal systolic array, and since the add-multiply cell is the basic PE in these circuits and the TPE cell is the reversible GF counterpart of this fundamental add-multiply PE, the systolic arrays shown in Sections 2.2 – 2.4 (cf. Figures 3 – 5) can be constructed reversibly using the interconnection between TPEs as well. The following chapter introduces the m-ary quantum realizations and computations of the m-ary reversible systolic arrays that were introduced in this chapter.

Chapter 5

QUANTUM SYSTOLIC NETWORKS

Quantum computing (QC) is a method of computation that uses a dynamic process governed by the Schrödinger Equation (SE) [1,66]. The one-dimensional time-dependent SE (TDSE) is as follows:

$$-\frac{(h/2\pi)^2}{2m}\frac{\partial^2 |\psi\rangle}{\partial x^2} + V|\psi\rangle = i(h/2\pi)\frac{\partial |\psi\rangle}{\partial t}, \qquad (36)$$

$$\text{or} \quad H|\psi\rangle = i(h/2\pi)\frac{\partial |\psi\rangle}{\partial t}, \qquad (37)$$

where h is Planck's constant ($6.626 \cdot 10^{-34}$ $J \cdot S$), $V(x, t)$ is the potential, m is particle's mass, i is the imaginary number, $|\psi(x,t)\rangle$ is the quantum state, H is the Hamiltonian operator $H = -[(h/2\pi)^2/2m]\nabla^2 + V$, and ∇^2 is the Laplacian operator.

A general solution to TDSE is the expansion of a stationary (i.e., time-independent or spatial) basis functions (i.e., eigen states) $u_e(\vec{r})$ using a time-dependent (i.e., temporal) expansion coefficients $c_e(t)$ as follows:

$$\psi(\vec{r},t) = \sum_{e=0}^{n} c_e(t) u_e(\vec{r}) =$$
$$c_0(t)u_0(\vec{r}) + c_1(t)u_1(\vec{r}) + c_2(t)u_2(\vec{r}) + \ldots + c_n(t)u_n(\vec{r}).$$

The expansion coefficients $c_e(t)$ are a scaled complex exponentials as follows:

$$c_e(t) = k_e e^{-i\frac{E_e}{(h/2\pi)}t},$$

where E_e are the energy levels. While the above holds for all physical systems, in the quantum computing context, the time-independent SE (TISE) is normally used:

$$\nabla^2 \psi = \frac{2m}{(h/2\pi)^2}(V - E)\psi, \qquad (38)$$

where the solution $|\psi\rangle$ is an expansion over orthogonal basis states $|\phi_i\rangle$ defined in a linear complex vector space called Hilbert space **H** as follows:

$$|\psi\rangle = \sum_i c_i |\phi_i\rangle, \qquad (39)$$

where the coefficients c_i are called probability amplitudes, and $|c_i|^2$ is the probability that the quantum state $|\psi\rangle$ will collapse into the (eigen) state $|\phi_i\rangle$. The probability is equal to the inner product $|\langle \phi_i | \psi \rangle|^2$, with the unitary condition $\sum |c_i|^2 = 1$. In QC, a linear and unitary operator ℑ is used to transform an input vector of quantum bits (qubits) into an output vector of qubits. In a two-valued QC, a qubit is a vector of bits defined as follows:

$$\text{qubit\_0} \equiv |0\rangle = \begin{bmatrix} 1 \\ 0 \end{bmatrix}, \text{qubit\_1} \equiv |1\rangle = \begin{bmatrix} 0 \\ 1 \end{bmatrix}. \qquad (40)$$

A two-valued quantum state $|\psi\rangle$ is a superposition of quantum basis states $|\phi_i\rangle$, (e.g., Equation (40)). Thus, for the orthonormal computational basis states $\{|0\rangle, |1\rangle\}$, one has the following quantum state:

$$|\psi\rangle = \alpha|0\rangle + \beta|1\rangle, \qquad (41)$$

where $\alpha\alpha^* = |\alpha|^2 = p_0 \equiv$ the probability of having state $|\psi\rangle$ in state $|0\rangle$, $\beta\beta^* = |\beta|^2 = p_1 \equiv$ the probability of having state $|\psi\rangle$ in state $|1\rangle$, and $|\alpha|^2 + |\beta|^2 = 1$. The calculation in QC for multiple systems (e.g., the equivalent of a register) follow the tensor product ($\otimes$). For example, given two states $|\psi_1\rangle$ and $|\psi_2\rangle$ one has the following:

$$|\psi_1\psi_2\rangle = |\psi_1\rangle \otimes |\psi_2\rangle, \qquad (42)$$
$$= (\alpha_1|0\rangle + \beta_1|1\rangle) \otimes (\alpha_2|0\rangle + \beta_2|1\rangle) = \alpha_1\alpha_2|00\rangle + \alpha_1\beta_2|01\rangle + \beta_1\alpha_2|10\rangle + \beta_1\beta_2|11\rangle.$$

A physical system, describable by the following Equation:

$$|\psi\rangle = c_1|\text{Spinup}\rangle + c_2|\text{Spindown}\rangle, \qquad (43)$$

(e.g., the hydrogen atom), can be used to physically implement a two-valued QC. Another common alternative form of Equation (43) is:

$$|\psi\rangle = c_1 \left|+\frac{1}{2}\right\rangle + c_2 \left|-\frac{1}{2}\right\rangle. \qquad (44)$$

Many-valued QC can also be accomplished. For the three-valued QC, the qubit becomes a 3-dimensional vector quantum discrete digit (qudit), and in general, for many-valued QC (MVQC) the qudit is of dimension "many". For example, one has for 3-state QC (in Hilbert space H) the following qudits:

$$\text{qudit}\_0 \equiv |0\rangle = \begin{bmatrix} 1 \\ 0 \\ 0 \end{bmatrix}, \text{qudit}\_1 \equiv |1\rangle = \begin{bmatrix} 0 \\ 1 \\ 0 \end{bmatrix}, \text{qudit}\_2 \equiv |2\rangle = \begin{bmatrix} 0 \\ 0 \\ 1 \end{bmatrix} \qquad (45)$$

A three-valued quantum state is a superposition of three quantum orthonormal basis states (i.e., vectors). Thus, for the orthonormal computational basis states $\{|0\rangle, |1\rangle, |2\rangle\}$, one has the following quantum state:

$$|\psi\rangle = \alpha|0\rangle + \beta|1\rangle + \gamma|2\rangle,$$

where $\alpha\alpha^* = |\alpha|^2 = p_0 \equiv$ the probability of having state $|\psi\rangle$ in state $|0\rangle$, $\beta\beta^* = |\beta|^2 = p_1 \equiv$ the probability of having state $|\psi\rangle$ in state $|1\rangle$, $\gamma\gamma^* = |\gamma|^2 = p_2 \equiv$ the probability of having state $|\psi\rangle$ in state $|2\rangle$, and $|\alpha|^2 + |\beta|^2 + |\gamma|^2 = 1$. The calculation in QC for many-valued multiple systems follow the tensor product in a manner similar to the one demonstrated for the higher-dimensional qubit in two-valued QC.

Many of the two-valued and multiple-valued quantum circuit implementations use two-valued and multiple-valued quantum Swap-based and Not-based gates [1]. This can be important, since the Swap and Not gates are basic primitives in quantum computing, from which many other gates are built, such as: (1) two-valued Not gate, (2) two-valued Controlled-Not gate (i.e., Feynman gate), (3) two-valued Controlled-Controlled-Not gate (i.e., Toffoli gate), (4) two-valued Swap gate, (5) two-valued Controlled-Swap gate (i.e., Fredkin gate), (6) multiple-valued Not gate, (7) multiple-valued Controlled-Not gate (i.e., multiple-valued Feynman gate), (8) multiple-valued Controlled-Controlled-Not gate (i.e., multiple-valued Toffoli gate), (9) multiple-valued Swap gate, and (10) multiple-valued Controlled-Swap gate (i.e., multiple-valued Fredkin gate).

Figure 16 shows important Galois-based two-valued and many-valued (i.e., m-ary) quantum primitives [1,66]. It has been shown that a physical system comprising trapped ions under multiple-laser excitations can be utilized to reliably implement many-valued quantum computing (e.g., Figure 17d) [65].

A physical system in which an atom (i.e., particle) is exposed to a specific potential field (i.e., potential function) can also be used to implement MVQC (where two-valued being a special case) (cf. Figure 17a) [1,66]. In such an implementation, the resulting distinct energy states are used as the orthonormal basis states.

Various physical realization methodologies for the implementation of two-valued and many-valued quantum computing have been presented [1,65,66]: (a) energy states for a simple harmonic oscillator potential for two-valued and many-valued quantum computing (cf. Figure 17a), (b) particle spin (i.e., angular momentum) for two-valued quantum computing (cf. Figure 17b), (c) light polarization for two-valued quantum computing (cf. Figure 17c), (d) cold trapped ions for two-valued and many-valued quantum computing (cf. Figure 17d), and (e) the quantum interference experiment (cf. Figure 17e), which by analogy to quantum computing implies that when a unitary quantum operator is applied, some amplitudes increase while others decrease (i.e., they interfere with each other), where the probability density function PDF = $P_{12} = |\beta|^2 = \beta\beta^*$ and $P_{12} = \gamma_1 P_1 + \gamma_2 P_2$.

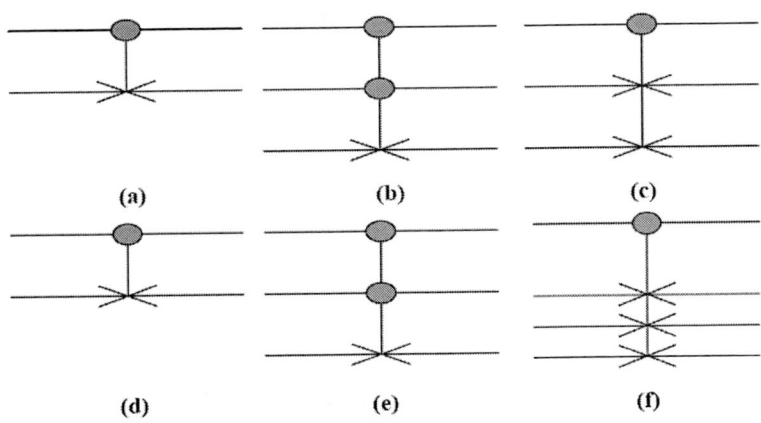

Figure 16. Schematics of basic quantum primitives: (a) binary Feynman (also called Controlled-Not: CN), (b) binary Toffoli (also called Controlled-Controlled-Not: C-C-NOT or C^2NOT), (c) binary Fredkin (also called Controlled-Swap: C-Swap), (d) ternary Feynman, (e) ternary Toffoli, and (f) ternary Fredkin.

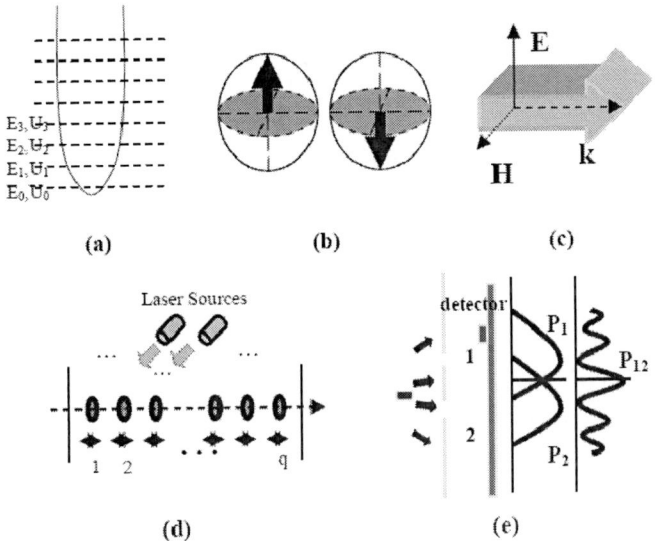

Figure 17. Technologies for various quantum realizations.

Similarly to the reversibility introduced in the previous chapter, and since basic PEs used in the construction of arithmetic systolic arrays are the add-multiply cells, Figure 18a introduces the ternary (3, 3) quantum Toffoli gate that implements in the Galois quantum domain the classical Kung add-multiply cell (in Figure 18b), and Figure 18c implements the 2D quantum Kung systolic array over GF(3) by interconnecting the general GF(3) reversible Toffoli gates.

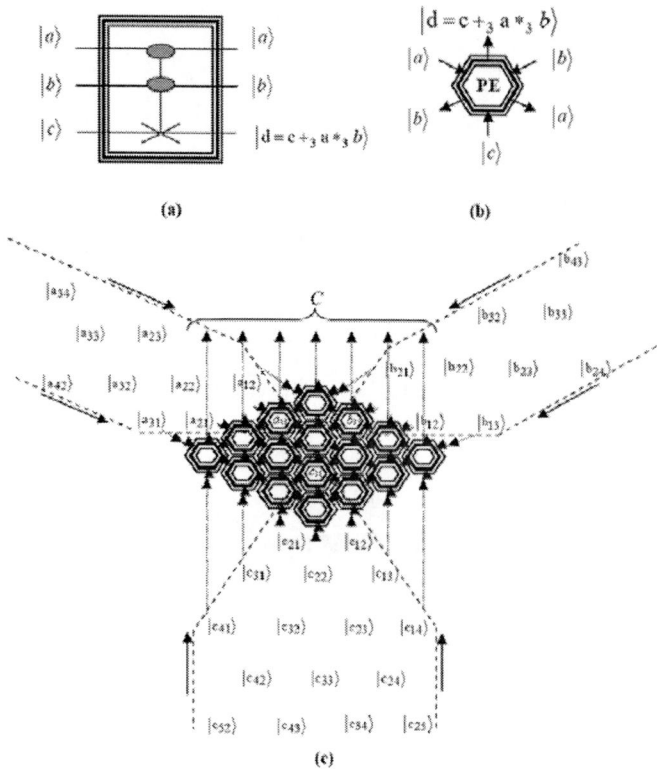

Figure 18. Third-radix Galois quantum Kung systolic network: (a) quantum (3, 3) Toffoli gate, (b) quantum (3, 3) Kung cell, and (c) quantum Kung systolic network.

Figure 19 shows the quantum realization for the reversible circuits in Figures 13a – 13c, respectively.

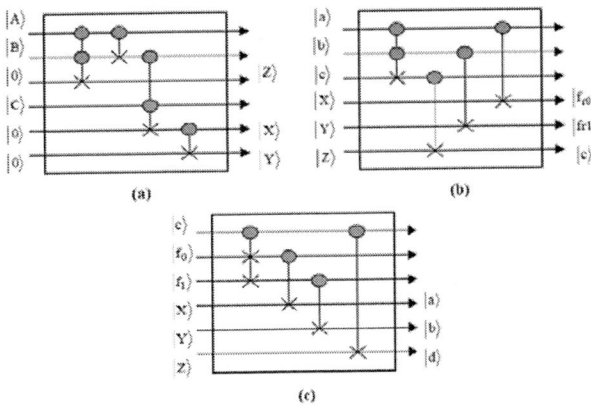

Figure 19. The corresponding quantum implementations for the reversible circuits in Figures 13a – 13c.

Figure 20 illustrates the unitary matrix representations of the following fundamental quantum gates [1]: (a) binary Feynman, (b) binary Swap, (c) binary Toffoli, (d) binary Fredkin, (e) ternary Feynman, and (f) ternary Swap. As will be shown, such matrix-based representations are used in the quantum computations that evolve the input qubits (and input qudits in GF(3)) to output qubits (and output qudits in GF(3)) through transformations that are implemented by the interconnection of various quantum gates.

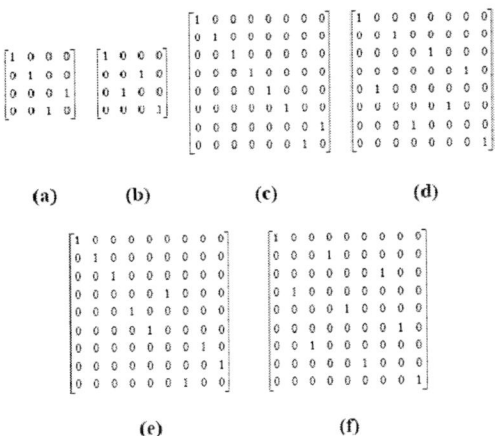

Figure 20. The unitary matrix representations for basic quantum primitives: (a) binary Feynman, (b) binary Swap, (c) binary Toffoli, (d) binary Fredkin, (e) ternary Feynman, and (f) ternary Swap.

Example 5. The evolution of quantum signals in Figure 18c is performed through the cascade (i.e., serial) evolution of input qubits using the cascading of serial binary Toffoli gates (cf. Figures 18a and 18b). Figure 21 shows the transformation of input qubits into output qubits through a single Toffoli cell (i.e., TPE).

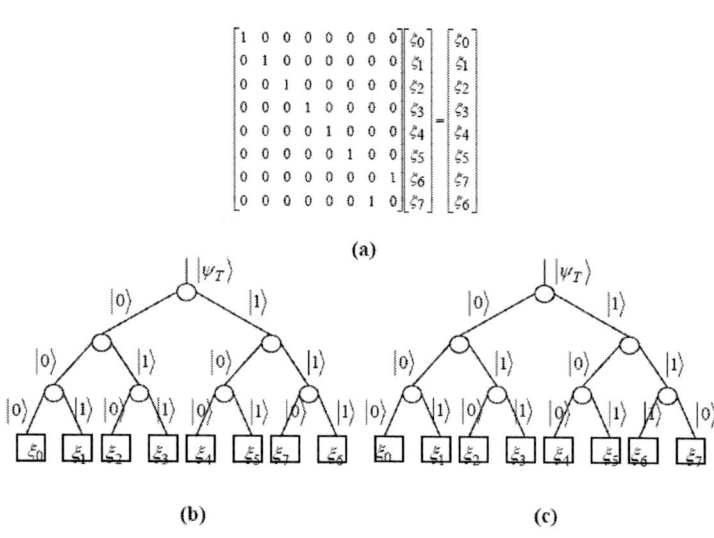

Figure 21. Representations using three-level binary Quantum Decision Trees (QDTs) for the quantum evolution of qubits through the Toffoli cell (in Figures 18a – 18b): (a) quantum evolution of input qubit using the Toffoli cell, (b) transformation of the probabilities' vector in the corresponding quantum space, and (c) the transformed quantum space (i.e., the transformed orthonormal axes) upon the input probabilities' vector where $\|\vec{\xi}\| = 1$.

While Example 5 considered that the probability of the quantum system is 100% in one state and 0% in the rest (i.e., $|\psi_i\rangle = 1.0|0\rangle + 0.0|1\rangle$ or $|\psi_i\rangle = 0.0|0\rangle + 1.0|1\rangle$), the following example illustrates the evolution of a general superimposed quantum state.

Example 6. Feynman gate is the quantum XOR and plays a fundamental role in quantum computing [1,7,66]. The fundamental Swap gate (cf. Figure 20b) can be decomposed into serially-interconnected Feynman-based primitives. This example illustrates the evolution of the input superimposed quantum states into output quantum states using the quantum circuit in Figure 22d which is related to the fundamental Swap gate via using the quantum circuit in Figure 22c. The quantum matrix representations for the basic gates in Figures 22a and 22b is shown within the two Figures respectively [1,66]. The quantum matrix representation for the circuit in Figure 22d is obtained using the regular matrix multiplication of the matrix representations of the serially interconnected C-Not and flipped C-Not quantum primitives.

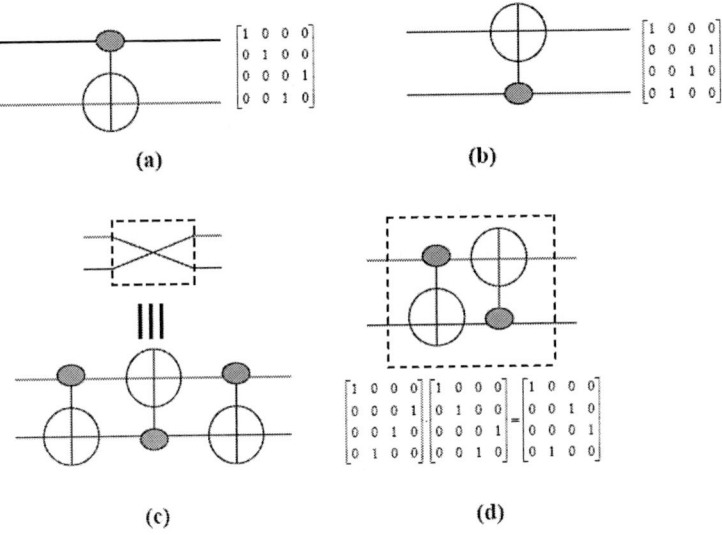

Figure 22. An example of quantum primitives and circuits and their quantum matrix representations: (a) 2-valued Controlled-Not gate (i.e., 2-valued C-Not gate or 2-valued Feynman gate), (b) flipped 2-valued Controlled-Not gate (i.e., flipped 2-valued C-Not gate or flipped 2-valued Feynman gate), (c) Swap gate as an equivalent to a serial cascading of CNot-FlippedCNot-CNot gates, and (d) a sub-circuit of the quantum circuit in Figure 22c which is composed of a serial interconnection between a C-Not gate and a flipped C-Not gate.

For the two quantum input states $|\psi_1\rangle = \frac{1}{\sqrt{2}}(|0\rangle + |1\rangle)$ and $|\psi_2\rangle = a|0\rangle + b|1\rangle$ the evolution of the superimposed input quantum state

$$|\psi\rangle = |\psi_1\psi_2\rangle = |\psi_1\rangle \otimes |\psi_2\rangle = \frac{1}{\sqrt{2}}(|0\rangle + |1\rangle) \otimes (a|0\rangle + b|1\rangle) = \frac{1}{\sqrt{2}}(a|00\rangle + b|01\rangle + a|10\rangle + b|11\rangle)$$

using the circuit in Figure 22d is obtained as

$$\begin{bmatrix} 1 & 0 & 0 & 0 \\ 0 & 0 & 1 & 0 \\ 0 & 0 & 0 & 1 \\ 0 & 1 & 0 & 0 \end{bmatrix} \begin{bmatrix} a/\sqrt{2} \\ b/\sqrt{2} \\ a/\sqrt{2} \\ b/\sqrt{2} \end{bmatrix} = \begin{bmatrix} a/\sqrt{2} \\ a/\sqrt{2} \\ b/\sqrt{2} \\ b/\sqrt{2} \end{bmatrix}.$$ Thus, the superimposed output state is produced as:

$$|\psi'\rangle = \frac{a}{\sqrt{2}}|00\rangle + \frac{a}{\sqrt{2}}|01\rangle + \frac{b}{\sqrt{2}}|10\rangle + \frac{b}{\sqrt{2}}|11\rangle = (a|0\rangle + b|1\rangle)\frac{1}{\sqrt{2}}(|0\rangle + |1\rangle) = |\psi_2\rangle \otimes |\psi_1\rangle = |\psi_2\psi_1\rangle.$$

One can note from Example 6 that the quantum matrix holds the orthonormal axes in the corresponding quantum space that define that particular quantum space, the quantum state is a vector (of probabilities) in that quantum space, and the quantum evolution is a transformation of the probabilities' vector in the corresponding quantum space. Equivalently, one may interpret the quantum evolution as the application of the transformed quantum space (i.e., the transformed orthonormal axes) upon the input probabilities' vector.

These two interpretations can be directly observed from the following Equation:

$$|\psi'\rangle = [|00\rangle \quad |01\rangle \quad |10\rangle \quad |11\rangle] \begin{bmatrix} 1 & 0 & 0 & 0 \\ 0 & 0 & 0 & 1 \\ 0 & 0 & 1 & 0 \\ 0 & 1 & 0 & 0 \end{bmatrix} \begin{bmatrix} 1 & 0 & 0 & 0 \\ 0 & 1 & 0 & 0 \\ 0 & 0 & 0 & 1 \\ 0 & 0 & 1 & 0 \end{bmatrix} \begin{bmatrix} a/\sqrt{2} \\ b/\sqrt{2} \\ a/\sqrt{2} \\ b/\sqrt{2} \end{bmatrix}$$

$$= [|00\rangle \quad |01\rangle \quad |10\rangle \quad |11\rangle] \begin{bmatrix} 1 & 0 & 0 & 0 \\ 0 & 0 & 1 & 0 \\ 0 & 0 & 0 & 1 \\ 0 & 1 & 0 & 0 \end{bmatrix} \begin{bmatrix} a/\sqrt{2} \\ b/\sqrt{2} \\ a/\sqrt{2} \\ b/\sqrt{2} \end{bmatrix} = [|00\rangle \quad |01\rangle \quad |10\rangle \quad |11\rangle] \begin{bmatrix} a/\sqrt{2} \\ a/\sqrt{2} \\ b/\sqrt{2} \\ b/\sqrt{2} \end{bmatrix}$$

$$= [|00\rangle \quad |11\rangle \quad |01\rangle \quad |10\rangle] \begin{bmatrix} a/\sqrt{2} \\ b/\sqrt{2} \\ a/\sqrt{2} \\ b/\sqrt{2} \end{bmatrix} = \frac{1}{\sqrt{2}}(a|00\rangle + b|11\rangle + a|01\rangle + b|10\rangle)$$

Although Example 6 demonstrates the method for evolving the input superimposed quantum states into output quantum states for two-valued quantum circuit, the same method is straightforward extended to the many-valued case by using the corresponding many-valued quantum matrix representations [1] (e.g., Figures 20e - 20f). The state in the general case of m-valued quantum Toffoli-based Kung systolic array can be either: (1)

decomposable into the tensor product as shown in Equation (46) for n quantum Toffoli gates (i.e., n quantum TPEs), or (2) non-decomposable (i.e., entangled) as shown in Equation (47) for n quantum Toffoli gates (i.e., n quantum TPEs).

$$|\psi_T\rangle = |\psi_{12...n}\rangle = \prod_{T=1}^{n}\left(\sum_{k=0}^{7}\xi_{kT}|D_{(67)}\rangle\right), \qquad (46)$$

$$|\psi_T\rangle = |\psi_{12...n}\rangle \neq \prod_{T=1}^{n}\left(\sum_{k=0}^{7}\xi_{kT}|D_{(67)}\rangle\right), \qquad (47)$$

where ξ_{kT} is the probability amplitudes and $|D_{(67)}\rangle$ is the group-theoretic representation of the binary Toffoli-transformed elementary basis states. The $|D_{(67)}\rangle$ group-theoretic representation of the binary Toffoli gate (cf. Figure 7b) stems from the fact that the TPE cell transforms the input qubit 6 into qubit 7 and input qubit 7 into qubit 6 while preserving the domain-range mapping for the rest of qubits from 0 to 5, i.e., $0 \to 0, 1 \to 1, 2 \to 2, 3 \to 3, 4 \to 4,$ and $5 \to 5$.

As an example of binary quantum superposition, Figure 23 shows a binary superimposed QDT where n is the number of binary quantum TPEs, and D_k is a level branch. Due to the fact that, in QC, all systems' states can occur at the same time (e.g., photon spin-up and spin-down in the two-valued QC), all of the tree paths from the root to the leaves in Figure 23 can occur simultaneously (i.e., in parallel), and only after measurement a single path will be observed as the whole system's composite (i.e., superimposed or correlated) state will collapse into that single path (i.e., state) after measurement. As an example, the observed path is shown in Figure 23 as a dark line. From computation point of view, each path in the Toffoli-based QDT in Figure 23 is a single calculation (i.e., computation or processing), and thus a massive computational parallelism occurs with 2^{3n} calculations performed simultaneously, and the QDT path superposition will collapse after measurement into a single path, where the path with the highest probability (i.e., leaf) α_j has the highest probability to be measured.

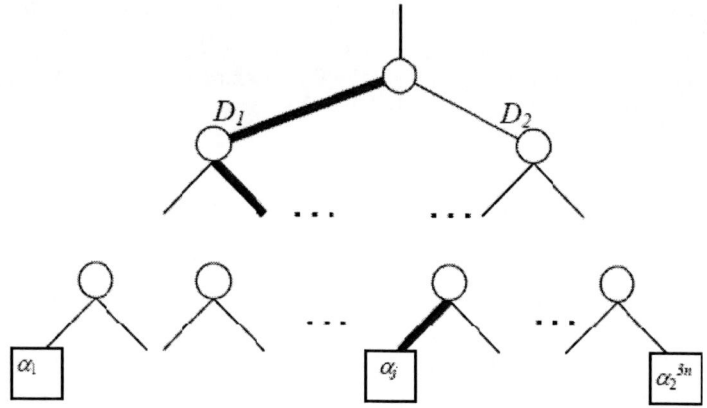

Figure 23. A superimposed n-stage TPE-based binary quantum decision tree.

As an example of entanglement in the context of two-valued quantum systolic networks, if the state vectors of certain quantum Toffoli cells (i.e., quantum TPEs) A and B in a quantum add-multiply-based systolic array were entangled with each other, then if one changes the state vector of one system A, the corresponding state vector of the other system B is also changed, instantaneously, and independently of the medium through which some communicating signal must travel. By measuring one of the state vectors of a quantum system A, the state vector collapses into a knowable state. Instantaneously and automatically, the state vector of the other quantum system B collapses into the other knowable state.

Chapter 6

CONCLUSIONS AND FUTURE WORK

Due to the quantum computing fact that all systems' states can exist at the same time, all of the states of the tree in quantum computing occur in parallel and only after measurement a single state will be observed as the whole system's superimposed state will collapse into that single state. From the processing perspective, each path in the tree within quantum computing is a single computation, and thus a massive computational parallelism exists with massive number of computations performed at the same time where the path with the highest probability has the highest probability to be measured.

New m-ary reversible and quantum systolic architectures for the synthesis of general m-ary Galois-based functions are introduced in this work. Since the architecture of conventional computers suffers from the two inherent difficulties of (a) long communication paths and (b) the fact that the single CPU sequentially fetches and executes instructions, systolic networks speed up the computation through (1) cost-effectiveness through simplicity and regularity, (2) concurrency (i.e., parallelism) through pipelining and multiprocessing, (3) regular communication wiring occurs only between neighboring processing elements (PEs) (i.e., the elimination of global broadcasting), and (4) I/O bandwidth-throughput improvements. In a systolic architecture, three factors are fundamental: (a) the type of the PE, (b) the systolic topology, and (c) the input/output ordering of data items in the I/O streams. The new arrays maintain the important regularity of the classical systolic arrays with the addition of preserving the reversibility property in the reversible space and the superposition property in the quantum space. Since it is estimated that at least 75% of all scientific applications involve some form of matrix calculations which are expensive in terms of storage space and processing time, and since basic PEs that are used in the construction of arithmetic systolic arrays are the add-multiply cells, the results introduced in this research are general and apply to a wide class of add-multiply-based systolic arrays and other types of systolic arrays as well.

Since the reduction of power consumption is a major requirement for the circuit design in several future technologies, the main features of several future technologies will include reversibility. Consequently, the new design method introduced in this research can play an important role in the design of circuits that consume minimal power for future circuit applications, and the new quantum superposition property will be essential in performing super-fast arithmetic-intensive computations that are basic in several matrix-based future implementations such as in multi-dimensional quantum signal processing.

The hardware implementation of the introduced m-ary reversible systolic architectures can be performed utilizing low-power technologies such as using the existing low-power VLSI adiabatic CMOS technology or using existing optical technology. On the other direction, the quantum PEs hardware realization of the introduced m-ary quantum systolic systems need to be further investigated; potential technologies for the quantum systolic hardware realization will include using the entangled particles' spin (i.e., angular momentum) for two-valued quantum systolic computing and cold trapped ions for the general case of many-valued quantum systolic computing. Important issues of the physical implementation for the quantum systolic data propagation and synchronization will need to be further investigated as well.

Future investigations will include the following items: (1) the investigation of implementing m-ary functions in 3D systolic networks; (2) the investigation of realizing the introduced quantum systolic networks using the quantum dot (QD) technology; (3) the investigation of increasing the power of reversible and quantum systolic networks by chaining several reversible and quantum systolic networks together; (4) the investigation of implementing reversible programmable systolic networks; (5) the investigation of creating new Galois reversible systolic networks to implement the Galois-based operations of addition, inversion, division, multiplication, exponentiations, solutions to linear systems of equations, and $C + AB^2$; (6) the investigation of the efficiency of using new types of reversible and quantum systolic networks for data coding and encryption applications; (7) the implementation of Faddeev's algorithm, which is a general purpose algorithm which is useful for a wide spectrum of matrix operations, using new reversible and quantum systolic networks; (8) the investigation of the issue of global quantum data clocking (i.e., synchronization); (9) numerical performance evaluation of the potential generalization of reversible systolic networks beyond simple reversible cellular systems; and (10) the investigation of the implementation of the new reversible and quantum systolic networks within the context of super-systolic arrays.

REFERENCES

[1] A. N. Al-Rabadi, Reversible Logic Synthesis: From Fundamentals to Quantum Computing (Springer-Verlag, New York, 2004).

[2] A. N. Al-Rabadi, Reversible Systolic Arrays, Part I: Two-Valued Bijective Single-Instruction Multiple-Data (SIMD) Architectures and Their Quantum Extensions, CD Proc. of the International Workshop on Spectral Methods and Multirate Signal Processing (SMMSP), (Moscow, Russia, September 1-2, 2007).

[3] A. N. Al-Rabadi, Reversible Systolic Arrays, Part II: m-Ary Equipollent SIMD Circuits and Their Quantum Realizations, CD Proc. of the International Workshop on Spectral Methods and Multirate Signal Processing (SMMSP), (Moscow, Russia, September 1-2, 2007).

[4] A. N. Al-Rabadi, Multiple-Level Circuit Solutions to the Circuit Non-Decomposability Problem of the Set-Theoretic Modified Reconstructability Analysis (MRA), *International Journal of General Systems (IJGS)*. 35, 2 (2006) 169 - 189.

[5] M. Annaratone et al., Architecture of Warp, Proc. Compcon, (IEEE CS Press, Los Alamitos, California, Order No. 764, February 1987), pp. 264 - 267.

[6] K. Araki, I. Fujita, and M. Morisue, Fast Inverters over Finite Field Based on Euclid's Algorithm, *Trans. IEICE.* 72E, 11 (November 1989) 1230 - 1234.

[7] A. Barenco, C. H. Bennett, R. Cleve, D. P. DiVincenzo, N. Margolus, P. Shor, T. Sleator, J. Smolin, and H. Weinfurter, Elementary Gates for Quantum Computation, *Phys. Rev. A.* 52 (1995) 3457 - 3467.

[8] G. H. Barnes, R. M. Brown, K. Maso, D. J. Kuck, D. L. Slotnick, and R. A. Stokes, The ILLIAC IV Computer, *IEEE Trans. Comp.* C-17 (1968) 746 - 757.
[9] C. Bennett, Logical Reversibility of Computation, *IBM J. of Research and Development.* 17 (1973) 525 - 532.
[10] E. R. Berlekamp, Algebraic Coding Theory (McGraw-Hill, New York, 1968).
[11] E. R. Berlekamp, G. Seroussi, and P. Tong, A Hypersystolic Reed-Solomon Decoder, Reed-Solomon Codes and their Applications, (S. B. Wicker and V. K. Bhargava, eds. Piscataway, New Jersey, IEEE Press, 1994), Chapter 10.
[12] R. E. Blahut, Theory and Practice of Error-Control Codes (Addison-Wesley, 1983).
[13] T. R. Blakeslee, Digital Design with standard MSI and LSI (John Wiley & Sons, New York, 1975).
[14] S. Borkar et al., iWarp: An Integrated Solution to High Speed Parallel Computing, Proc. Supercomputing, Vol. 1 (CS Press, Los Alamitos, California, Order No. 882, 1988), pp. 330 - 339.
[15] R. P. Brent and H. T. Kung, Systolic VLSI Arrays for Polynomial GCD Computation, *IEEE Trans. Comp.* 13 (1984) 731 - 736.
[16] H. Brunner, A. Curiger, and M. Hofstetter, On Computing Multiplicative Inverses in $GF(2^m)$, *IEEE Trans. Comp.* 42, 8 (August 1993) 1010 - 1015.
[17] D. E. Denning, Cryptography and Data Security (Addison-Wesley, 1983).
[18] A. De Vos, Reversible Computing, *Progress in Quantum Electronics.* 23 (1999) 1 - 49.
[19] W. Diffie and M. E. Hellman, New Directions in Cryptography, *IEEE Trans. Information Theory.* 22 (1976) 644 - 654.
[20] A. L. Fisher, K. Sarocky, and H. T. Kung, Experience with the CMU Programmable Systolic Chip, Proc. Soc. of Photo-Optical Instrumentation Engineers, Real Time Signal Processing VII (SPIE, San Diego, California, 1984), p. 495.
[21] J. Fortes and B. W. Wah, Systolic Arrays - From Concept to Implementation, *IEEE Computer Magazine.* 20 (July 1987) 12 - 17.
[22] D. Foulser and R. Scheiber, The Saxpy Matrix-1: A General-Purpose Systolic Computer, *Computer.* 20, 7 (July 1987) 35 - 43.
[23] E. Fredkin and T. Toffoli, Conservative Logic, *International Journal of Theoretical Physics.* 21 (1982) 219 - 253.

References

[24] B. Friedlander, Block Processing on a Programmable Systolic Array, Proc. Int. Conf. Parallel Processing (IEEE CS Press, Los Alamitos, California, Order No. 889, 1987), pp. 184 - 187.

[25] P. Frison et al., Micmacs: A VLSI Programmable Systolic Architecture, in Systolic Array Processors, (Prentice-Hall, Englewood Cliffs, New Jersey, 1989), pp. 145 - 155.

[26] W. K. Fuchs and E. E. Swartzlander Jr., Wafer-Scale Integration: Architectures and Algorithms, *Computer.* 25, 4 (April 1992) 6 - 8.

[27] P. Greussay, Programmation des Mega-Processeurs: du GAPPala Connection Machine, Course Notes DEA Artificial Intelligence (Paris University, VIII, Vincennes, 1985).

[28] M. Gokhale et al., Splash: A Reconfigurable Linear Logic Array, Proc. Int. Conf. Parallel Processing (IEEE CS Press, Los Alamitos, California, Order No. 2101, 1990), pp. 1526 - 1531.

[29] M. Gokhale et al., Building and Using a Highly Parallel Programmable Logic Array, *Computer.* 24, 1 (January 1991) 81 - 89.

[30] L. K. Grover, A Fast Quantum-Mechanical Algorithm for Database Search, Proc. Symp. On Theory of Computing (STOC) (1996), pp. 212 - 219.

[31] J.-H. Guo and C-L. Wang, Systolic Array Implementation of Euclid's Algorithm for Inversion and Division in $GF(2^m)$, *IEEE Trans. Comp.* 47, 10 (October 1998) 1161 - 1167.

[32] I.-H. Guo and C.-L. Wang, Hardware-Efficient Systolic Architecture for Inversion and Division in $GF(2^m)$, *IEEE Trans. Computers and Digital Techniques.* (1998) 272 - 278.

[33] S. Hammering, A Note on Modifications to the Givens Plane Rotation, *J. Ins. Math. Appl.* 13 (1974) 215 - 218.

[34] M. A. Hasan, Double-Basis Multiplicative Inversion over $GF(2^m)$, *IEEE Trans. Comp.* 47, 9 (September 1998) 960 - 970.

[35] M. A. Hasan and V. K. Bhargava, Bit-Serial Systolic Divider and Multiplier for Finite Fields $GF(2^m)$, *IEEE Trans. Comp.* 41, 8 (August 1992) 972 - 980.

[36] C.A.R. Hoare, Communicating Sequential Processes, *Communications of the ACM.* 21 (1978) 666 - 677.

[37] C.-T. Huang and C.-W. Wu, High-Speed C-Testable Systolic Array Design for Galois-Field Inversion, Proc. European Design and Test Conference (March 1997), pp. 342 - 346.

[38] R. Hughey and D. Lopresti, Architecture of a Programmable Systolic Array, Proc. Int. Conf. Systolic Arrays (IEEE CS Press, Los Alamitos, California, Order No. 860, 1988), pp. 41 - 49.

[39] R. Hughey and D. Lopresti, B-SYS: A 470-Processor Programmable Systolic Array, Proc. Int. Conf. Parallel Processing (IEEE CS Press, Los Alamitos, Calif., Order No. 2355-22, 1991), pp. 1580-1583.
[40] K. Hwang, Computer Arithmetic: Principles, Architecture, and Design (John Wiley & Sons, 1979).
[41] K. Hwang and Y. H. Cheng, Partitioned Matrix Algorithms for VLSI Arithmetic Systems, *IEEE Trans. Comp.* C-31, 12 (December 1982) 1215 - 1224.
[42] K. Hwang and F. Briggs, Computer Architecture and Parallel Processing (McGraw-Hill, 1984).
[43] M. Ishii et al., Cellular Array Processor CAP and Application, Proc. Int. Conf. Systolic Arrays (IEEE CS Press, Los Alamitos, California, Order No. 860, 1988), pp. 535 - 544.
[44] A. K. Jain, Fundamentals of Digital Image Processing (Prentice-Hall, 1989).
[45] S. Jones, A. Spray, and A. Ling, A Flexible Building Block for the Construction of Processor Arrays, in Systolic Array Processors (Prentice-Hall, Englewood Cliffs, N. J., 1989), pp. 459-466.
[46] R. M. Kant and T. Kimura, Decentralized Parallel Algorithms for Matrix Computation, Proc. of the 5th Annual Symposium on Computer Architecture (Palo Alto, California, April 1978), pp. 96 - 100.
[47] T. Kean and J. Gray, Configurable Hardware: Two Case Studies of Micrograin Computation, in Systolic Array Processors (Prentice-Hall, Englewood Cliffs, N. J., 1989), pp. 310-319.
[48] C. K. Ko and O. Wing, Mapping Strategy for Automated Design of Systolic Arrays, Proc. lnt. Conf. Systolic Arrays (IEEE CS Press, Los Alamitos, Calif., Order No. 860, 1988), pp. 285 - 294.
[49] A. Krikelis and R. M. Lea, Architectural Constructs for Cost-Effective Parallel Computers, in Systolic Array Processors (Prentice-Hall, Englewood Cliffs, N. J., 1989), pp. 287 - 300.
[50] D. J. Kuck, ILLIAC IV Software and Application Programming, *IEEE Trans. Comp.* C-17 (1968) 758 - 770.
[51] D. J. Kuck, A Survey of Parallel Machine Organization and Programming, *Computing Surveys.* 9 (1977) 29 - 59.
[52] H. T. Kung, Why Systolic Architectures? *Computer.* 15, 1 (January 1982) 37 - 46.
[53] S. Y. Kung, VLSI Array Processors (Prentice-Hall, 1988).
[54] H. T. Kung and C. E. Leiserson, Systolic Array Apparatuses for Matrix Computations, (U. S. Patent Application, Filed December 1978).

References

[55] H. T. Kung, C.-C. Wang, C.-J. Huang, and K.-C. Tsai, A 1.00 GHz 0.6-µm 8-Bit Carry Lookahead Adder Using PLA-Styled All-n-Transistor Logic, *IEEE Trans. Circuits and Systems-II: Analog and Digital Signal Processing.* 47 (2000) 133 - 135.

[56] K. T. Johnson, A. R. Hurson, and B. Shirazi, General-Purpose Systolic Arrays, *Computer.* 26, 11 (November 1993) 20 - 31.

[57] R. Landauer, Irreversibility and Heat Generation in the Computational Process, *IBM Journal of Research and Development.* 5 (1961) 183 - 191.

[58] D. Landis et al., A Wafer-Scale Programmable Systolic Data Processor, Proc. 9th Biennial University/Government/Industry Microelectronics Symposium (IEEE Press, Piscataway, N. J., 1991), pp. 252 - 256.

[59] R. M. Lea, The ASP: A Fault-Tolerant VLSI/ULSI/WSI Associative String Processor for Cost-Effective Processing, Proc. Int. Conf. Systolic Arrays, pp. 515 - 524.

[60] P. Lenders and H. Schroder, A Programmable Systolic Device for Image Processing Based on Morphology, *Parallel Computing.* 13, 3 (March 1990) 337 - 344.

[61] B. Lindscog and P. E. Danielsson, PICAP3: A Parallel Processor Tuned for 3D Image Operations, Proc. 8th Int. Conf. Pattern Recognition (IEEE CS Press, Order No. 742, (microfiche only), 1986), pp. 1248 - 1250.

[62] M. Malek and E. Opper, The Cylindrical Banyan Multicomputer: A Reconfigurable Systolic Architecture, *Parallel Computing.* 10, 3 (May 1989) 319 - 326.

[63] E. D. Mastrovito, VLSI Architectures for Computations in Galois Fields (Ph.D. Thesis, Linkoping University, 1991).

[64] C. A. Mead and L. A. Conway, Introduction to VLSI Systems (1978).

[65] A. Muthukrishnan and C. R. Stroud, Multi-Valued Logic Gates for Quantum Computation, *Phys. Rev. A.* 62 (2000) 052309.

[66] M. Nielsen and I. L. Chuang, Quantum Computation and Quantum Information (Cambridge University Press, 2000).

[67] A. V. Oppenheim and R. W. Schafer, Digital Signal Processing (Prentice-Hall, 1975).

[68] A. V. Oppenheim and R. W. Schafer, Discrete-Time Signal Processing (Prentice-Hall, 1999).

[69] C. Paar, Some Remarks on Efficient Inversion in Finite Fields, Proc. Int. Symp. Information Theory (1995).

[70] C. Paar, Fast Inversion in Composite Galois Fields $GF(2^m)$, Proc. Int. Symp. Information Theory (1998).
[71] K. K. Parhi, VLSI Digital Signal Processing Systems (New York, John Wiley & Sons, 1999).
[72] P. Picton, Optoelectronic Multi-Valued Conservative Logic, *Int. J. of Optical Computing.* 2 (1991) 19 - 29.
[73] K. W. Przytula and J. B. Nash, Implementation of Synthetic Aperture Radar Algorithms on a Systolic/Cellular Architecture, Proc. Int. Conf. Systolic Arrays (IEEE CS Press, Los Alamitos, California, Order No. 860, 1988), pp. 21 - 27.
[74] P. Quinton and Y. Robert, Systolic Algorithms & Architectures (Prentice-Hall, N. J., 1991).
[75] L. R. Rabiner and R. W. Schafer, Digital processing of Speech Signals (Prentice-Hall, 1987).
[76] I. S. Reed and T. K. Truong, The Use of Finite Fields to Compute Convolutions, *IEEE Trans. Information Theory.* 21 (March 1975) 208 - 213.
[77] S. M. Reddy, Easily Testable Realizations of Logic Functions, *IEEE Trans. Comp.* C-21 (1972) 1183 - 1188.
[78] K. Roy and S. Prasad, Low-Power CMOS VLSI Circuit Design (John Wiley & Sons Inc., 2000).
[79] A. H. Sameh and D. J. Kuck, On Stable Parallel Linear System Solvers, *Journal of the Association for Computing Machinery* (ACM). 25 (1978) 81 - 91.
[80] D. V. Sarwate and N. R. Shanbhag, High-Speed Architectures for Reed-Solomon Decoders, *IEEE Trans. VLSI Systems.* 9, 5 (October 2001) 941 - 955.
[81] H. Schroder and P. Strazdins, Program Compression on the ISA, *Parallel Computing.* 17, 2-3 (June 1991) 207 - 215.
[82] P. W. Shor, Algorithms for Quantum Computation: Discrete Logarithms and Factoring, Proc. Symp. Foundations of Computer Science (1994), pp. 124 - 134.
[83] R. Smith and G. Sobelman, Simulation-Based Design of Programmable Systolic Arrays, *Computer-Aided Design.* 23, 10 (December 1991) 669 - 675.
[84] L. Snyder, Introduction to the Configurable Highly Parallel Computer, *Computer.* 15, 1 (January 1982) 47 - 56.
[85] R. S. Stankovic, Spectral Transform Decision Diagrams in Simple Questions and Simple Answers (Nauka, Belgrade, 1998).

[86] H. S. Stone, Parallel Computations, in Introduction to Computer Architecture, ed. H. S. Stone (Science Research Associates, Chicago, 1975), pp. 318 - 374.

[87] I. E. Sutherland and C. A. Mead, Microeleclronics and Computer Science, *Scientific American.* 237 (1977) 210 - 228.

[88] N. Takagi, A VLSI Algorithm for Modular Division Based on the Binary GCD Algorithm, *IEICE Trans. Fundamentals of Electronics, Comm., and Computer Sciences.* E81-A, 5 (May 1998) 724 - 728.

[89] K. J. Thurber and L. D. Wald, Associative and Parallel Processors, *Computing Surveys.* 7 (1975) 215 - 255.

[90] M. Toverud and V. Anderson, CESAR: A Programmable High-Performance Systolic Array Processor, Proc. Int. Conf. Computer Design (IEEE CS Press, Los Alamitos, California, Order No. 872, (microfiche only), 1988), pp. 414 - 417.

[91] C.-L. Wang and J.-H. Guo, New Systolic Arrays for $C + AB^2$, Inversion, and Division in $GF(2^m)$, *IEEE Trans. Comp.* 49, 10 (October 2000) 1120 - 1125.

[92] C.-C. Wang, P.-M. Lee, R.-C. Lee, and C.-J. Huang, A 1.25 GHz 32-Bit Tree-Structured Carry Lookahead Adder, Proc. Int. Symp. Circuits and Systems (ISCAS), 4 (2001), pp. 80 - 83.

[93] Y. Watanabe, N. Takagi, and K. Takagi, A VLSI Algorithm for Division in $GF(2^m)$ Based on Extended Binary GCD Algorithm, *lEICE Trans. Fundamentals of Electronics, Comm., and Computer Sciences.* E85-A, 5 (May 2002) 994 - 999.

[94] S.-W. Wei, VLSI Architectures for Computing Exponentiations, Multiplicative Inverses, and Divisions in $GF(2^m)$, Proc. Int. Symp. Circuits and Systems (ISCAS) (1994), pp. 203-206.

[95] S.-W. Wei, VLSI Architectures for Computing Exponentiations, Multiplicative Inverses, and Divisions in $GF(2^m)$, *IEEE Trans. Circuits and Systems: Analog and Digital Signal Processing.* 44, 10 (October 1997) 847 - 855.

[96] C. Wenyang, L. Yanda, and J. Yue, Systolic Realization for 2D Convolution Using Configurable Functional Method in VLSI Parallel Array Designs, *Proc. IEEE, Computers and Digital Technology.* 138, 5 (September 1991) 361 - 370.

[97] C. H. Wu, C. M. Wu, M. D. Shieh, and Y. T. Wang, Systolic VLSI Realization of a Novel Iterative Division Algorithm over $GF(2^m)$: A High-Speed, Low-Complexity Design, Proc. Int. Symp. Circuits and Systems (ISCAS) (2001), pp. 33 - 36.

[98] C. H. Wu, C. M. Wu, M. D. Shieh, and Y. T. Wang, An Area-Efficient Systolic Division Circuit over $GF(2^m)$ for Secure Communication, Proc. Int. Symp. Circuits and Systems (ISCAS) (2002), pp. 733 - 736.
[99] Z. Yan and D. Y. Sarwate, Systolic Architectures for Finite Field Inversion and Division, Proc. Int. Symp. Circuits and Systems (ISCAS) (2002), pp. 789 - 792.
[100] Z. Yan and D. V. Sarwate, New Systolic Architectures for Inversion and Division in $GF(2^m)$, *IEEE Trans. Comp.* 52, 11 (2003) 1514 - 1519.

INDEX

A

advantages, 3, 6
algorithm, 9, 12, 28, 29, 30, 48
amplitudes, 36, 38, 45
architecture, x, 11, 12, 47
arithmetic, vii, x, 6, 7, 25, 40, 48

B

background, 3
bandwidth, 6, 7, 9, 12, 47
basis functions, 35
binary, 22
binary logic, 22
breakdown, 1
building blocks, 1

C

CAP, 54
coding, x, 48
coefficients, 35, 36
communication, vii, 2, 6, 47
communication paths, 2, 47
complexity, 28
computation, vii, 1, 2, 5, 6, 12, 15, 28, 35, 45, 47
computing, vii, x, 1, 2, 5, 35, 38, 47, 48
Computing, 53
concurrency, 47
configuration, 21
configurations, 6
Conservative, 52
consumption, vii, 2, 48
Controlled-Controlled-NOT, 38
Controlled-NOT, 38, 43
correlation, 9
cost, vii, 6, 47
CPU, 47

D

damages, iv
decomposition, 2, 7
DFT, 5, 6, 9
distributed computing, x

E

efficiency, 48
encryption, 48
engineering, 7, 11
equipment, 5
expansion, 35, 36

F

Feynman gate, 38, 43
FFT, 9
formal language, 6
Fredkin gate, 38
function values, 21

G

garbage, 15, 28
gate, 38
gates, 22, 38
graph, 6

H

Hamiltonian, 35
Hilbert space, 36, 37
host, 5
hybrid, 30
hydrogen, 37

I

ideal, 2
image, x
integration, 6
interface, 7
interference, 38
inversion, 5, 6, 7, 48
ions, 38, 48

J

Jordan, ix

L

light, 38
linear systems, 2, 5, 48

M

manufacturing, 28
mapping, 15, 25, 28, 29, 30, 33, 45
matrix, 2, 5, 6, 7, 9, 11, 12, 18, 19, 20, 21, 23, 25, 26, 27, 28, 32, 41, 43, 44, 47, 48
memory, 5, 6
methodology, 19
modules, 6
momentum, 38, 48
Moscow, 51
multiple-valued, 38
multiplication, 2, 5, 7, 9, 11, 12, 17, 28, 30, 31, 32, 43, 48
multiplier, 31

N

nanotechnology, x

O

operator, 31, 38
overlap, 2

P

parallel, vii, x, 1, 2, 5, 30, 45, 47
parallel algorithm, 2
parallelism, vii, 1, 2, 3, 45, 47
performance, vii, 2, 6, 49
polarization, 38
ports, 6, 7
probability, vii, 1, 36, 37, 38, 39, 42, 45, 47
probability density function, 39
programming, 6
propagation, 48
pumps, 5

Q

quantum bits, 36
quantum computing, vii, x, 1, 36, 38, 42, 47
quantum dot, 48
quantum state, 35, 36, 37, 42, 43, 44
qubits, 36, 41, 42, 45

R

realization, 38
recognition, 6
recommendations, iv
respect, 17
response time, 6
robotics, x
Russia, 51

S

Shannon, 22
signals, 42
space, 2, 19, 36, 42, 44, 47
spin, 38, 45, 48
storage, 47
streams, 7, 8, 47
substitution, 18, 25, 31
synchronization, 48, 49
synthesis, 2, 25, 28, 47

T

testing, 17
Toffoli gate, 38
topology, 47
transformation, 42, 44
transformations, 9, 11, 41
trapped ions, 38, 48
two-valued, 38
two-valued Controlled-Swap gate, 38
two-valued Swap gate, 38

U

uniform, 6
unitary, 38

V

vector, 1, 2, 5, 9, 11, 12, 15, 19, 32, 36, 37, 42, 44, 46